Springer Series in Materials Science

Volume 351

The Springer Series in Materials Science covers the complete spectrum of materials research and technology, including fundamental principles, physical properties, materials theory and design. Recognizing the increasing importance of materials science in future device technologies, the book titles in this series reflect the state-of-the-art in understanding and controlling the structure and properties of all important classes of materials.

Maurizio Vedani

Metal Science in Modern Manufacturing Technologies

Springer

Maurizio Vedani
Department of Mechanical Engineering
Politecnico di Milano
Milan, Italy

ISSN 0933-033X ISSN 2196-2812 (electronic)
Springer Series in Materials Science
ISBN 978-3-031-97124-2 ISBN 978-3-031-97125-9 (eBook)
https://doi.org/10.1007/978-3-031-97125-9

This Springer imprint is published by the registered company Springer Nature Switzerland AG
The registered company address is: Gewerbestrasse 11, 6330 Cham, Switzerland

Foreword

The main goal of this work is to highlight the relations of material composition and processing technologies on the resulting microstructure and product performance. This cross-correlation is widely cited in the literature and in the web, but it appears that a deep knowledge which considers all the involved scientific aspects is still lacking in the background of material specialists working in many industrial fields and of academic scholars. (Fig. 1).

The book targets professionals and researchers who already possess a sound background about the principles of physical metallurgy, including crystal structure, phase diagrams, diffusion phenomena, mechanical properties of metals and alloys.

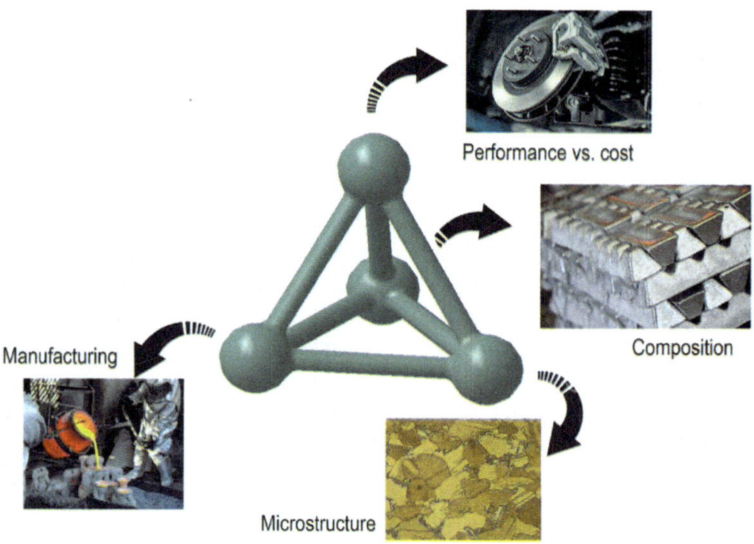

Fig. 1 Graphical view of the cross-correlation existing among material properties, manufacturing technology and product performance

It is believed that an improved theoretical knowledge about the physical laws ruling metal fabrication, microstructure development, and achievable properties of products would help solving many technical concerns of industrial experts and ease the understanding of material selection and property definition for the final applications. With this aim, the approach adopted for this book is such that the reader will mainly find information about the phenomena affecting metals and alloys during their manufacturing route and on how to govern them to improve the workability and tune the final properties of parts. The perspective is therefore on the material side, while concepts about the metalworking technologies will be provided only as case studies to support the understanding of the former view.

"Actually, we're looking for a metallurgist with less experience and more PhD."

To close this brief introduction, as a scholar working on fundamental and applied research for the development of metal alloys and of their processing routes, I always had in my office a cartoon, which I found in a journal at the times of my Ph.D., many years ago. Originally, I framed and kept the above image since it was funny, but along these decades it has been instructive in continuously recalling the importance of acquiring a clear practical experience on the processing steps of materials, that absolutely needs to be paired with a sound knowledge about the fundamental phenomena governed by physical metallurgy principles. As for many other facts of life, the truth lies in the middle. I am convinced that a sound synergy between "experience and Ph.D." can make the difference.

Milano (Italy)
July 2025

Maurizio Vedani

Competing Interests The author has no competing interests to declare that are relevant to the content of this manuscript.

Contents

Chapter 1
Solidification and Casting

The ability to cast metals is a heritage which dates back to the ancient times. The first historical eras following the stone age (before about 4000 B.C.) are classified based on the ability to cast different metals: the copper age, the bronze age, the iron age. The technical skills were even considered as strategic secrets to be jealously guarded, since the ability to manufacture efficient weapons and tools was of paramount importance for the life and safety of entire populations.

Thousands of years later, we are still exploiting the same concepts of melting and freezing alloys, but the awareness and consciousness by which we govern these processes have dramatically increased. But still there is room for improving our knowledge on fundamental principles, designing new casting alloys and making processes more efficient and sustainable.

1.1 Solidification of Metals and Alloys

Before considering any solidification mechanism, it is worth spending few words about the structure of liquid metals, as opposed to that of solid metals and of gases.

A solid crystalline metal features a regular arrangement of atoms according to an almost stable specific lattice configuration. Although crystal defects such as vacancies and dislocation play and important role, its ordered structure rules most of the typical metal properties. On the other end, in (ideal) gases the atoms are depicted as non-interacting particles placed at much larger interatomic distances, occupying random and dynamically-changing positions.

The study of liquid metals has been the subject of great efforts ever since the sixties of last century (Ashcroft, 1969). This challenging topic is complicated by difficulties encountered in experimental testing, also associated to the dynamic nature of the liquid state. It is known that liquid metals do not possess a long-range ordered structure but the fact that the density between liquid and solid phases of metal does

M. Vedani, *Metal Science in Modern Manufacturing Technologies*, Springer Series in Materials Science 351, https://doi.org/10.1007/978-3-031-97125-9_1

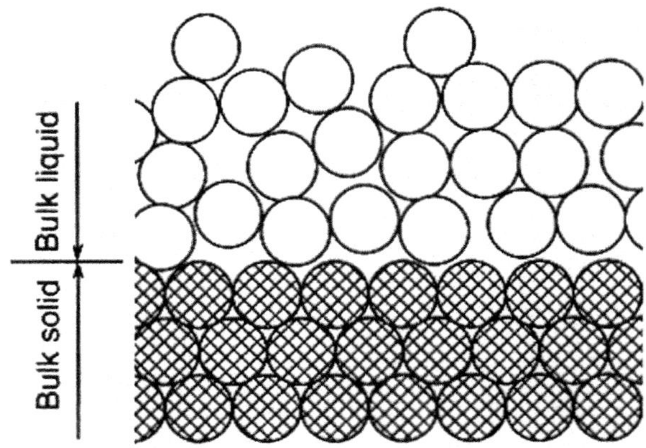

Fig. 1.1 Schematic view at the atomic scale of the structure of liquid and solid metallic phases across the solidification front

not differ significantly, suggests that the average interatomic distance between atoms is close to that of solids just below their melting point. Considering the relatively low amount of energy released upon melting (the latent heat of melting is of the order of 1/25 to 1/40 the latent heat of vaporization), it can be stated that also the bonding energy of atoms in the solid and liquid phases of metals are similar. X-ray diffraction studies allowed demonstrating that atoms in liquid metals might be arranged in an ordered structure in the short range, but the presence of a large amount of various defects prevents the achievement of a crystal configuration similar to that of solid metals. Finally, the diffusivity of atoms in liquids is orders of magnitude larger than that of solids owing to the low energy barrier for atomic movements. As a result, they show the typical fluidity, namely the inability to sustain any shear stress even of very low intensity (Abbaschian, 2009).

Given this configuration, the interface between a solid metal and its liquid counterpart (e.g. the solidification front while it is moving upon solidification) can be schematically depicted as shown in Fig. 1.1.

The movement of the interface separating the liquid and solid phases is macroscopically ruled by thermodynamic principles. If, for a given metal held at a fixed temperature, the internal energy of the solid phase is lower than that of the liquid phase (see Fig. 1.2a), the system will evolve promoting the transformation toward the solid phase and the solidification front will shift inside the liquid, replacing it with new volumes of the solid phase. At the atomic scale, this behaviour will correspond to the progressive deposition at the interface of atoms which leave the liquid phase and join the solid lattice. For sake of precision, we shall imagine that the flux of atoms crossing the interface is not exactly unidirectional. More precisely, an average rate of deposition of atoms at the solid front and a rate of detachment from it (proportional to the rate of freezing and of melting, respectively) can be defined. The qualitative trend of the freezing and melting rates against temperature are depicted in Fig. 1.2b. It can be readily understood that the crossing point of the two curves represents the equilibrium melting point. Moving to the right of the equilibrium melting point, makes the detachment of atoms from the solid phase predominant and the net rate of

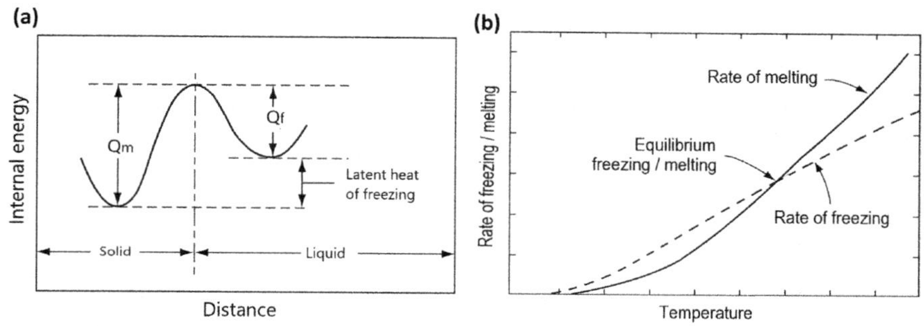

Fig. 1.2 Representation of the free energy gap governing the stability of the solid over the liquid phase (**a**) and trend of the freezing and melting rates versus temperature (**b**)

melting progressively increases while the opposite holds (the freezing rate increases) when lowering the temperature below the melting point (Chalmers, 1954).

It can also be deduced that the solidification rate can be significantly increased by promoting undercooling conditions, i.e. by rapidly placing a liquid metal far below its equilibrium freezing point due to the divergent nature of the two lines in Fig. 1.2b.

A further factor affecting the local solidification rate of a metal is the orientation of the crystal lattice exposed to the liquid. In simplified terms, the less close-packed the crystal plane, the higher the probability for a liquid atom to find a stable position and attach to the solid phase. This behaviour can be quantified through a specific parameter called the accommodation factor. As shown in Fig. 1.3, from simple geometrical considerations it can be realized that, for the same solidification conditions, the low atomic-density planes will statistically grow faster that the close-packed planes. Indeed, it is experimentally verified that the crystal directions leading to the fastest atom deposition are the $< 100 >$ for the face-centered and body-centered cubic cells, and the $< 10\overline{1}0 >$ for the hexagonal close-packed cells.

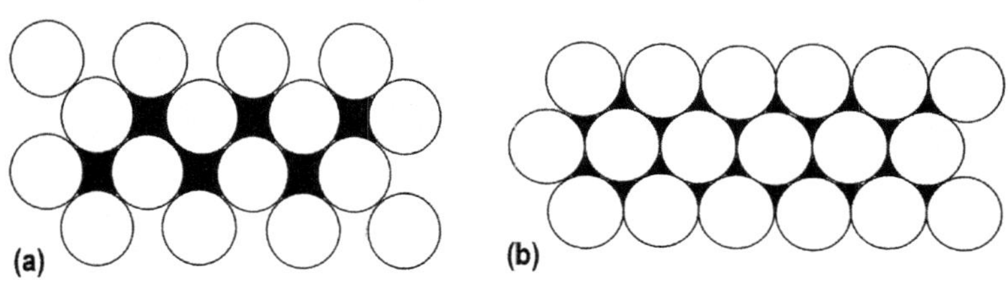

Fig. 1.3 The larger "pockets" left among atoms on the {100} plane of a face-centered cubic crystal (a) lead to a faster growth rate that the smaller gaps found on the {111} plane (b)

1.1.1 *Nucleation of a Solid from the Liquid Melt*

An additional issue to be considered for the nucleation of a solid phase from a liquid metal concerns the creation of the new solid–liquid interface. The actual stability of a new crystal volume will not only depend on the decrease in free energy corresponding to the phase transition, as previously described, but also on the surface-energy contribution associated to the growing interface. The net change in free energy per unit volume (ΔG) related to a solid cluster of atoms (assumed to be a sphere of radius r) can be expressed as the algebraic sum of the volume free energy (ΔG_V) and the surface free energy (ΔG_S):

$$\Delta G = \Delta G_V + \Delta G_S \approx \frac{-4}{3}\pi r^3 \frac{L\Delta T}{T_m} + 4\pi \gamma_{SL} r^2 \tag{1.1}$$

being L the latent heat of melting per unit volume, γ_{SL} the solid–liquid interface free energy and $\Delta T = T_m - T$ the amount of supercooling. When a melt is placed at $T < T_m$, the two contributions in Eq. 1.1 have opposite signs, with a trend against the radius of the sphere shown in Fig. 1.4. It can be observed that at large values of r, the cubic dependence of the volumetric term dominates over the surface term. Therefore, the net ΔG will go through a maximum at a specific value of r* and then it will assume a decreasing trend. This means that there exists a critical size below which the crystal (often called embryo) is unstable, despite the presence of a supercooling condition. On the contrary, when the value of the critical radius is exceeded, any additional atom joining the crystal will lower the free energy, therefore spontaneous crystal growth is promoted and the embryo becomes a stable nucleus.

The critical radius and the corresponding ΔG* can be readily evaluated by setting the condition d(ΔG)/dr = 0 in Eq. 1.1:

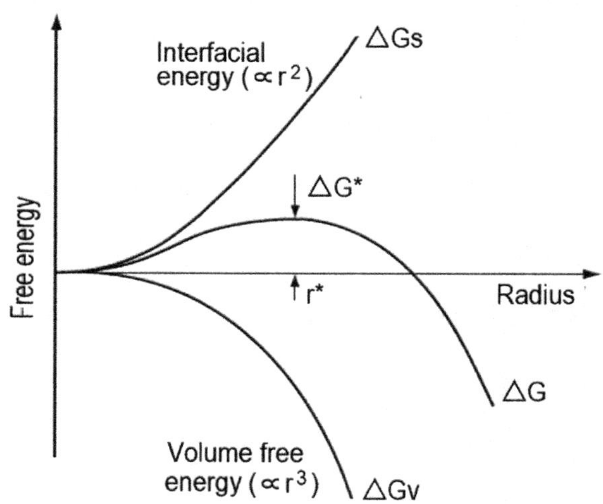

Fig. 1.4 The free energy of a solid sphere as a function of its radius r at a given supercooling ΔT

$$\frac{d(\Delta G)}{dr} = -4\pi r^2 \frac{L\Delta T}{T_m} + 8\pi \gamma_{SL} r = 0 \quad r^* = 2\gamma_{SL} \frac{T_m}{L\Delta T}$$

$$\Delta G^* = \frac{16}{3}\pi \frac{\gamma_{SL}^3 T_m^2}{L^2 \Delta T^2} \tag{1.2}$$

The described condition for the stable formation of a crystal within a homogeneous melt is called homogeneous nucleation. In a real environment such condition is hardly met since the probability that the statistical fluctuations of the atoms in a liquid can create a nucleus with a radius of size r* are very low. As a matter of fact, homogeneous nucleation is only possible with extensive supercooling of the liquid melt as demonstrated by the data collected in Table 1.1.

More frequent nucleation conditions are created in practical situations when heterogeneous solid particles in the melt or the same surface of the mould containing the melt catalyse the nucleation, stabilising nuclei at much lower supercooling and with a smaller size. The geometrical relationship between the wetting angle θ formed between an existing solid surface and the corresponding volume of the solid metal spherical cap of radius r (see Fig. 1.5) allows defining a modification of Eq. 1.2 for the expression of the critical ΔG* in case of heterogeneous nucleation (Laughlin & Hono, 2014):

$$\Delta G^* = \frac{16}{3}\pi \frac{\gamma_{SL}^3 T_m^2}{L^2 \Delta T^2} f(\theta) \quad \text{being} \quad f(\theta) = \frac{(2+\cos\theta)(1-\cos\theta)^2}{4} \tag{1.3}$$

The analytical expression of the factor f(θ) suggests that when θ = 180°, corresponding to the complete lack of wetting, f(θ) = 1 and the solid does not interact at all with the substrate. Hence, we are back to the ideal condition of homogeneous nucleation. On the opposite side, when θ = 0°, corresponding to perfect wetting, f(θ) = 0 and ΔG* = 0, which suggests the ideal occurrence of nucleation even without the need of any supercooling.

Table 1.1 Thermodynamic data and typical undercooling required to achieve homogeneous nucleation in some pure metals and in pure water

Material	Melting temperature T_m (°C)	Latent heat of melting L (J/kg)	Solid–liquid interface energy γ_{SL} (J/m^2)	Supercooling for homogeneous nucleation ΔT (°C)
Ga	30	$80 \cdot 10^3$	$56 \cdot 10^{-3}$	76
Cu	1085	$205 \cdot 10^3$	$177 \cdot 10^{-3}$	236
Ni	1453	$293 \cdot 10^3$	$255 \cdot 10^{-3}$	480
Fe	1538	$247 \cdot 10^3$	$204 \cdot 10^{-3}$	420
H_2O	0	$334 \cdot 10^3$	$44 \cdot 10^{-3}$	40

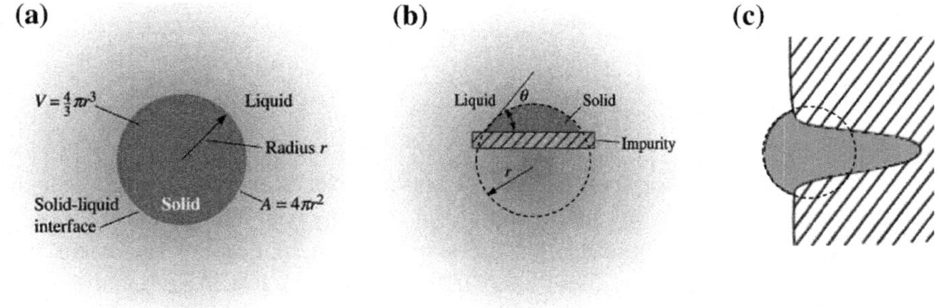

Fig. 1.5 The geometrical conditions ruling homogeneous (**a**) and heterogeneous nucleation on a flat surface (**b**) rather than from scratches or rough surfaces of the mould (**c**)

Figure 1.5 summarizes the different configurations described. Additionally, it is suggested that possible scratches or any other form of pores exposed on the surface of the substrate can reduce even more the actual ΔG^* value for nucleation.

1.1.2 Solidification of Pure Metals

After discussing the conditions for the nucleation of the solid phase from a liquid, we can now progress into the mechanisms of growth of the solid, first evaluating the simplified case of a pure metal. Under common solidification conditions, we can suppose that the liquid metal is poured into a mould and heat is mainly extracted from mould walls. We can study the evolution of the solid–liquid interface, the solidification front, initially growing as a macroscopically flat surface. At any stage of solidification, we can assume that the temperature gradient across the solidification interface is positive, the temperature monotonically rising when moving into the liquid, which is held at a temperature greater than the freezing point, as shown in Fig. 1.6.

Recalling the dependence of the melting/freezing rates from the gap to the equilibrium melting temperature (see Fig. 1.2b), it can be realized that any protuberance that by chance extends more than the average position of the interface into the liquid will be subjected to a hotter environment. The growth rate of this portion will thus slow down and the remainder of the interface will reach it again, restoring the flat surface. For similar reasons, any portion that by chance may lag behind the average position of the interface will accelerate its growth rate, restoring the original condition, Therefore, it can be concluded that such a positive temperature gradient in the liquid phase promotes a stable planar growth of the solidification front.

A different scenario is found when a liquid metal is poured into a chilled mould which induces a faster heat extraction, also leading to extensive liquid supercooling before the solid phase can form. Upon fast solidification, the latent heat released at the interface will be hardly distributed in the surrounding volume and will mainly

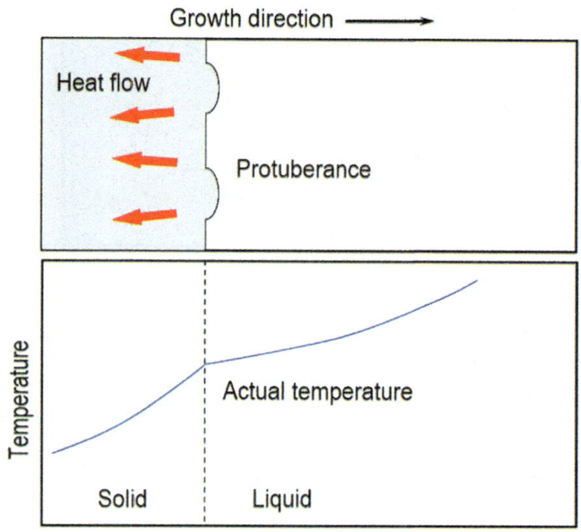

Fig. 1.6 The positive temperature gradient across the solid–liquid interface resulting in stable growth of the solidification front

concentrate into the liquid immediately facing the solidification front, raising the local temperature toward the freezing point, as depicted in Fig. 1.7.

Under these conditions the temperature profile drops, at least for a short distance, when moving into the liquid. As a result, any protuberance that by chance extends more than the average position of the interface into the liquid will be subjected to a cooler environment. This portion will be stimulated to grow further and faster and becomes unstable, generating peculiar structures called dendrites.

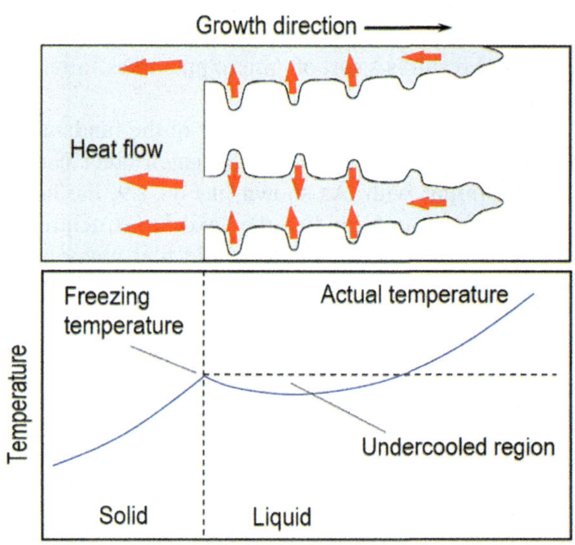

Fig. 1.7 The negative temperature gradient in the liquid facing the solid–liquid interface resulting in unstable dendritic growth of the solidification front

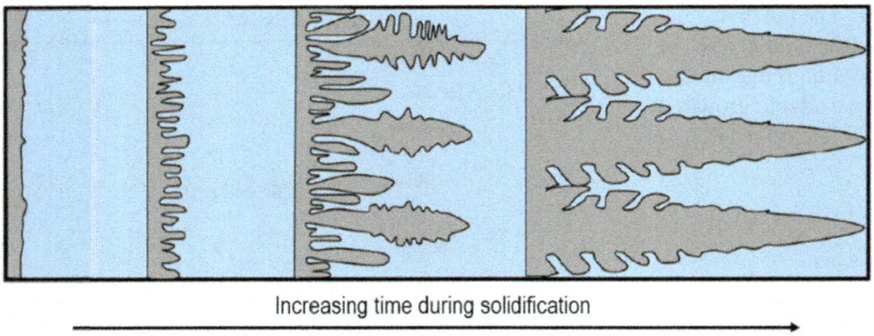

Fig. 1.8 Schematic of the development of an initial instability of the solidification front leading to a final steady state dendritic pattern with uniform spacing (adapted from Strickland et al., 2020)

Figure 1.8 schematically shows the complex shape and growth sequence of dendritic crystals according to the described mechanisms (Strickland et al., 2020). It must be considered that, associated with the growth of each dendrite, there is the release of a given amount of heat of fusion which locally raises the temperature in the liquid volume surrounding it delaying the growth of any similar protrusion within a short lateral distance. This eventually will generate almost equally spaced protrusions (primary branches of the dendrites) growing parallel to each other. Figure 1.8 also shows that the dendrites can grow developing additional secondary and even tertiary arms along direction perpendicular to each other. It can be speculated that, associated to the growth of a primary branch, there is also a lateral release of heat which again raises the temperature at the lateral solid–liquid interface of the dendrite with respect to the midway point between two adjacent dendrites. Therefore, this decreasing temperature gradient will again promote the unstable lateral growth of any protuberance along preferred crystal directions normal to those of the primary branches. Similarly, tertiary arms will form from the secondary branches and so on. The final result is a sort of "micro pine trees" after which the name derived from the Greek word "dèndro" for tree.

Finally, it is to remark that most of the fundamental phenomena described so far for metal solidification perfectly match other natural situations we could even be more familiar with. As shown in Fig. 1.9, the heterogeneous nucleation of water droplets from air humidity, the dendritic structure of ice and snow crystals are just few examples, but the reader could find many other phenomena in everyday life situations ruled by the same principles.

1.1.3 Solidification of Binary Alloys

When metallic alloys instead of pure metals are considered, further issues related to element distribution need to be added to the discussion. Here we will only introduce few simplified cases for binary alloys to evaluate how compositional effects are

Fig. 1.9 Examples of solidification mechanisms from nature. The heterogeneous nucleation of water droplets from vapour on a leaf substrate (**a**), the formation of dendritic ice crystals on a tree branch (**b**), the microscopic structure of a metal dendrite observed at the scanning electron microscope (**c**)

affected by the extent of the diffusional flow made possible by the cooling conditions experienced in real processes.

The starting situation refers to the ideal case of solidification under perfect equilibrium conditions, namely with associated full redistribution of the alloying elements in the solid and in the liquid phases, generating a homogeneous composition within each phase, at any stage of solidification. A further simplifying hypothesis implies the analysis of unidirectional solidification conditions, idealized by the planar growth of a solid–liquid front within a thermally insulated tube, where heat is extracted from one end only (the left end in the following Fig. 1.10) and the solidification front will grow rightward, consuming the liquid contained in the tube.

Let us consider the binary system depicted in the phase diagram of Fig. 1.10a, and an alloy of composition C_0. We shall follow the progress of the equilibrium solidification reading the actual compositions of the two phases (the liquid and the α-phase) on the horizontal tie-lines at any specific temperature. Correspondingly, the position of the solidification front and the actual composition of the two phases found in the sample tube filled with our alloy are plotted at three discrete time steps in Figs. 1.10b–d.

It is inferred that at a temperature infinitesimally below the liquidus value T_L, the very first layer of solid of composition $C_S = kC_0$ forms. The constant k represents the so-called partition coefficient (in general: $k = C_S/C_L$) which accounts for the different solubility of the solute atoms in the solid and liquid phases. As the temperature of the system is lowered and solidification progresses, both phases get enriched in the solute and the compositions of liquid and solid are kept perfectly homogeneous throughout their volumes by virtue of the hypothesis of complete diffusion. As expected, solidification ends at $T = T_S$, when the solid phase reaches the composition $C_S = C_0$ and the amount of liquid phase becomes nil.

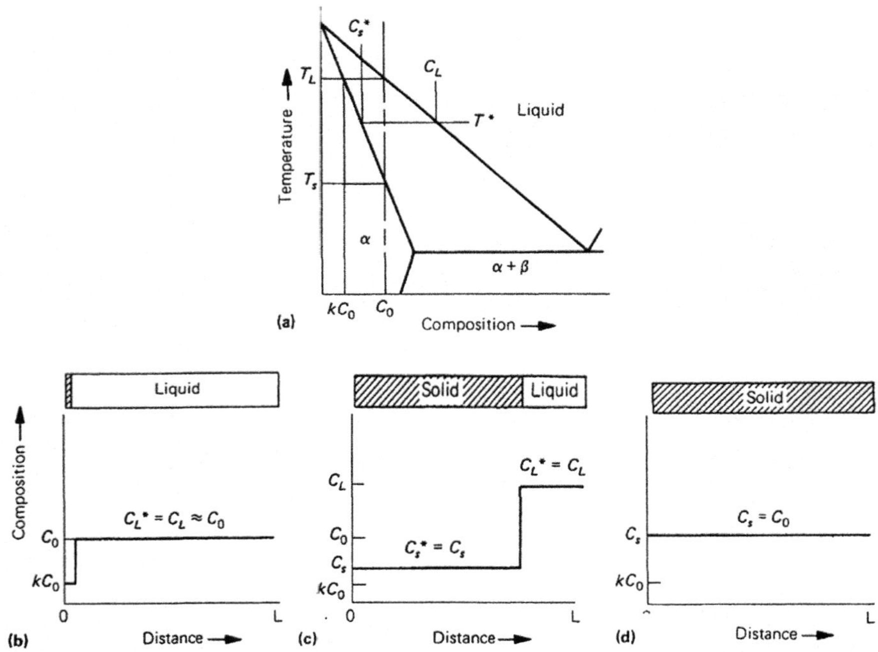

Fig. 1.10 Solute redistribution in an alloy of composition C_0 under equilibrium conditions following the phase diagram (**a**) and evolution of the composition of the liquid and solid phases during solidification (**b–d**)

Even though equilibrium phase diagrams can be useful for the prediction of the trends in solidification and for phase stability of metallic alloys, it is obvious that the underlying conditions almost never occur in nature nor in real technological processes. To approach real conditions, further assumptions have to be made. Consider now a second situation whereby we accept the idea of complete mixing of the solute atoms in the liquid phase due to faster intrinsic diffusion and the effects of convection. Moreover, the solute partitioning between the liquid and solid phases at the solidification interface again can still be ruled by equilibrium, but no diffusion at all is allowed in the solid phase. The evaluation of the solidification sequence under these assumptions is depicted in Fig. 1.11.

The main difference now stands in the compositional gradient created in the solid volume as the solidification progresses in absence of diffusion. As for the previous case, the first layer of solid created has a composition $C_S = kC_0$, but the compositional changes of the newly formed solid fractions at the distinct temperatures is kept visible by the gradient of solute of the solid phase (see Fig. 1.11c–d). A second important difference consists in the fact that solidification does not end anymore when the solid "locally" reaches the composition C_0 but at a later stage and at a lower temperature, specifically when the average composition in the whole solid phase (namely the integral of the curve in Fig. 1.11d averaged on the distance) equals C_0.

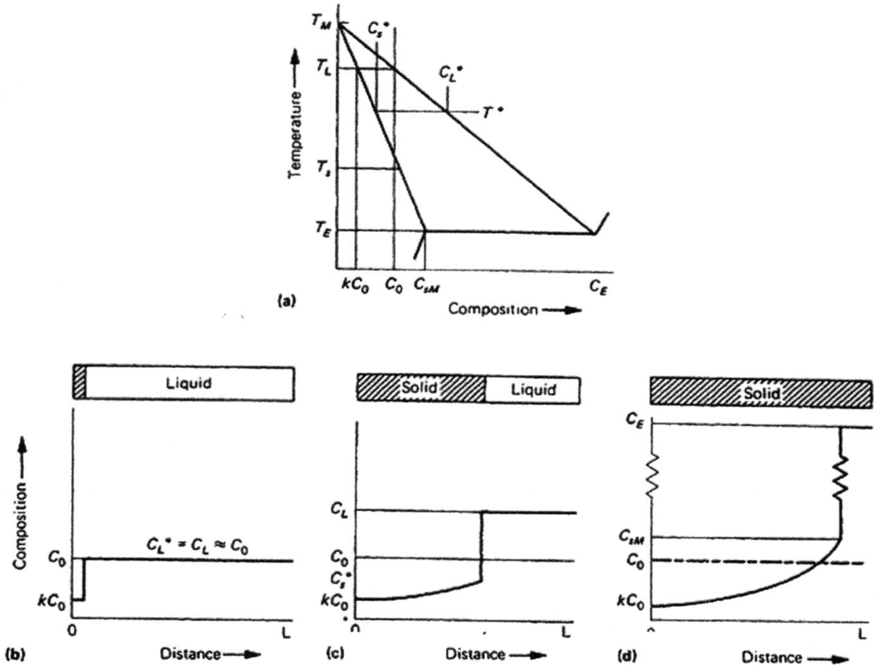

Fig. 1.11 Solute redistribution in an alloy of composition C_0 under conditions of absence of diffusion in the solid and full diffusion in the liquid phase, following the phase diagram (**a**) for partitioning of solute at the solidification interface and evolution of the composition of the liquid and solid phases during solidification (**b–d**)

Moreover, the specific example given in Fig. 1.11, suggests that when the solidification stage is significantly extended down to low temperatures, unexpected constituents may appear in the microstructure. This is the case of the eutectic formed at $T = T_E$ when the liquid reaches the composition $C_L = C_E$, which was not predicted based on the simple equilibrium diagram.

The solute concentration profile of the solid qualitatively depicted in Fig. 1.11 can be analytically estimated recalling the above-described hypotheses (complete diffusion in the liquid, no diffusion in the solid and equilibrium at the solidification interface) and assuming that the partition coefficient k is constant with temperature and for any cooling rate experienced upon solidification. By a balance of the amount of solute migrating from the liquid to the solid across the interface, it is possible to derive the analytical expression of the composition profile. According to the Scheil-Gulliver model, the solute composition in the solid phase (C_S) against the fraction of solid formed (f_S) is given by the following equation:

$$C_S = kC_0(1 - f_S)^{(k-1)} \tag{1.4}$$

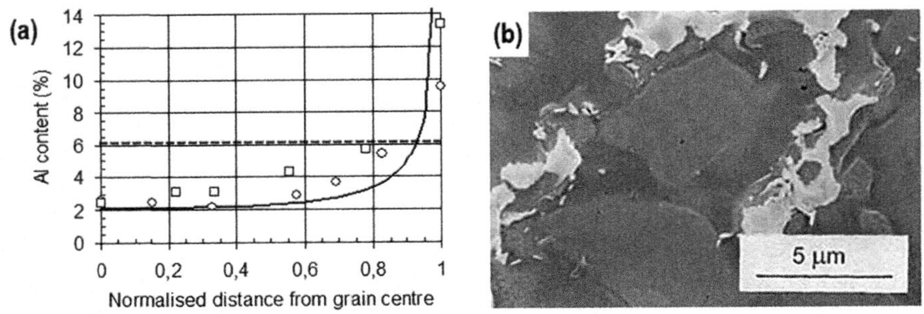

Fig. 1.12 Al concentration profile within the grains of a Mg-6wt.% Al diecast alloy (**a**) and corresponding microstructure revealing the formation of the eutectic constituent at the grain boundaries (**b**)

Figure 1.12 shows an example of application of the Scheil-Gulliver model to a diecast Mg-6%Al alloy solidified at a fairly high rate. In Fig. 1.12a the distinct datapoints refer to experimental measurements of the solute content versus the distance from grain center (the measurements were repeated for several grains having different sizes, hence the normalized distance) while the continuous line is the estimated solute profile calculated by Eq. 1.4. The experimental data show a good agreement with the modelled trend and allow realizing the extent of the fluctuations in solute concentration found in a real solidification process. A depletion of the Al content down to about 2% was measured at grain interiors (to be compared to the nominal alloy composition of 6%) while an enrichment exceeding 14% is achieved at the grain boundaries. In Fig. 1.12b the actual microstructure is represented, confirming the accumulation of Al at the grain boundaries, leading to the formation of the Al–$Al_{12}Mg_{17}$ eutectics.

The effect of the solute clustering in the last portions of the solidified structure is a typical phenomenon found in casting processes, known as elemental segregation. When limited to the short-range scale of the grains, it is referred to as micro-segregation, whereas when recorded at the macroscopic scale (e.g. in large billets, considering the accumulation of solute at their center) it is called macro-segregation.

An alternative way to illustrate the extent of the described segregation phenomena consists in evaluating the composition of the solid through the tie-lines of the phase diagram, selecting a number of discrete temperature steps, as shown in Fig. 1.13.

The discretization of the solidification process implies that the solid grains will be formed by the accumulation of distinct shells keeping their original composition as defined by the equilibrium hypotheses at the solid–liquid interface. The dashed line diverging from the equilibrium solidus line with decreasing temperature represents the average composition of the solid phase at a given stage of solidification (identified as nonequilibrium solidus temperature), it is calculated as the average solute concentration of all the already existing solid shells. The graph also reminds that solidification ends when the average composition of the solid phase reaches back the general composition of the alloy. Finally, it should be noted that in most real solidification processes, the extent of segregation is usually less than that predicted by the

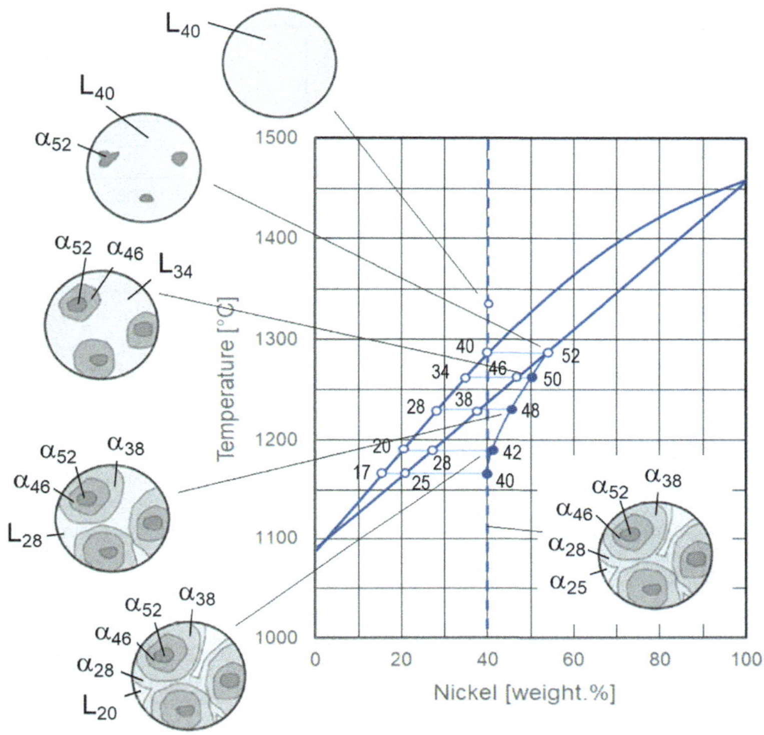

Fig. 1.13 Solute segregation formed upon solidification of a binary alloy, evaluated by the discretization of alloy shells which maintain their original composition set at the solidification interface, as defined by the phase diagram

Scheil-Gulliver equation or by this graphical method since solid state diffusion will promote solute redistribution, slightly reducing the sharp compositional gradient in the solid phase.

The third simplifying assumption we could remove to approach a more realistic solidification picture concerns the diffusion of the solute in the liquid. In the absence of stirring or significant convection flow in the melt, we can assume that the solute rejected into the liquid at the solid–liquid interface will be transported away only by diffusion. Although the diffusivity in liquid metals is much faster than in solids, it is reasonable to accept the idea that perfect redistribution of the solute will not be achieved during real solidification conditions. Considering the initial stages of unidirectional solidification of a binary alloy with the phase diagram of Fig. 1.14a, the first solid to form will be depleted in solute atoms, which will accumulate at the opposite side of the solidification front, in the liquid volume (see Fig. 1.14b). As more solid is formed, the amount of solute at the liquid side will increase until a steady state condition is attained (Fig. 1.14c). This occurs when the liquid next to the interface reaches a composition $C_L = C_0/k$ which enables the formation of a solid

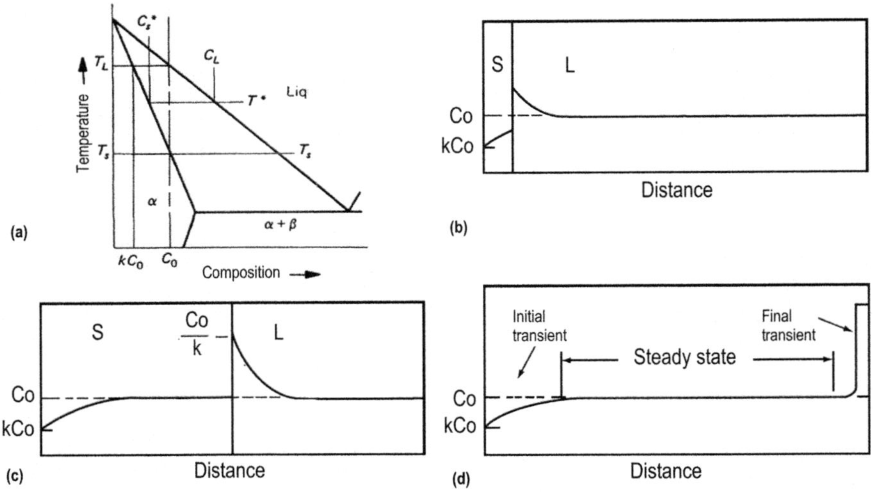

Fig. 1.14 Evolution of the composition profile across the solid–liquid interface for the alloy shown in the phase diagram (**a**) at three different stages of solidification (**b–d**), assuming finite diffusion in the liquid phase

with the composition C_0. In other terms, the solid formed from the enriched liquid layer facing the interface has the same solute content as the liquid drawn into the enriched layer. This condition is often referred to as dynamic equilibrium. For sake of completeness, in the last stages of solidification shown in Fig. 1.14d, the excess of solute atoms will accumulate in a closed volume without the ability to be rejected farther and will result in a final rise of the concentration profile, which compensates for the starting solute-depleted region.

Both the above treated situations used to evaluate the non-equilibrium solidification conditions can consistently explain phenomena such as the segregation of solute elements within the microstructure and the extension of the solidification range to lower temperatures than the equilibrium. A further aspect implying the instability of the solidification front will now be discussed in relation to the solute enrichment in the liquid layer facing the solidification front. For a pure metal, it has already been shown that the instability of the front might be induced by thermal supercooling effects, when the liquid melt is exposed to a negative temperature gradient in front of the interface, promoting a faster growth of any protuberance which can extend more in the liquid.

During the solidification of an alloy, a further condition can trigger the instability of the solidification front. Recalling that in most practical casting processes the hot metal is poured into a mould and freezes as a result of the outward heat flow through the mould walls, one can assume that a positive temperature gradient is set within the liquid, i.e. the temperature is lowest at the solidification front and rises toward the centre of the liquid volume, preventing any significant thermal supercooling.

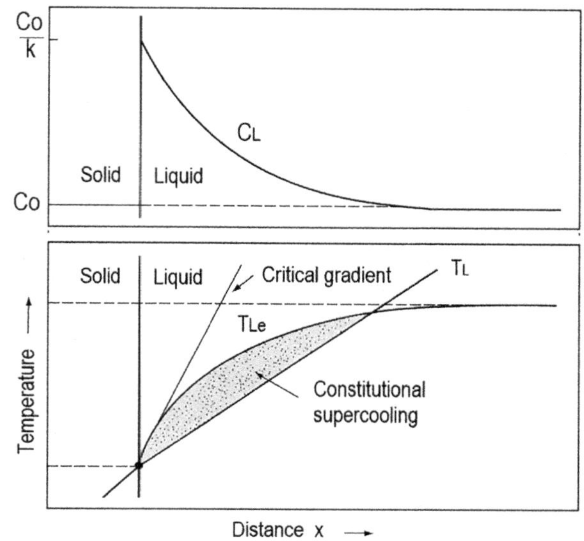

Fig. 1.15 The profile of the equilibrium temperature of a binary alloy resulting from the accumulation of solute ahead of the solidification front as compared to the actual temperature measured in the liquid. The shaded area in the lower plot represents the distance over which supercooling is created

Assuming a finite degree of diffusion in the liquid and the absence of convection or stirring, it is expected that a solute-enriched layer is formed in front of the interface with a specific composition profile ruled by the alloy considered, the solidification rate and the heat extraction efficiency. Fig. 1.15 qualitatively recalls the shape of a possible solute profile in the liquid field in terms of C_L values against the distance from the interface x. Referring now to the related alloy phase diagram, for each specific solute concentration value, an equilibrium solidification temperature T_{Le} can be derived and it can again be plotted against the distance from the interface, as shown in the bottom part of Fig. 1.15. A further line is reported in the plot labelled as T_L, namely the actual temperature, featuring the expected positive gradient, measured in the liquid.

Interestingly, the comparison of the two curves highlights that there can be a region ahead of the solid–liquid interface where the actual temperature of the liquid is lower than the equilibrium liquidus temperature, generating a new condition for the supercooling, now related to the liquid composition (or constitution), hence the name of constitutional supercooling (Messler, 1999).

For a given profile of the solute distribution, when the actual temperature gradient is progressively lowered, an initially planar front can become unstable and, when falling below the critical gradient depicted in Fig. 1.15, it can lead to the breakdown of the interface into a cellular structure. The cells will grow parallel to the direction of heat flow and according to the preferred crystal direction. When a sufficiently low gradient is reached, secondary arms are observed to grow laterally from the cells (or primary arms) giving rise to a dendritic structures. Figure 1.16 summarizes the various solidification growth modes expected in metallic alloys as a function of the extent of compositional supercooling created in the liquid. It is also reported that, when the supercooling distance exceeds a limit, new freestanding grains can nucleate

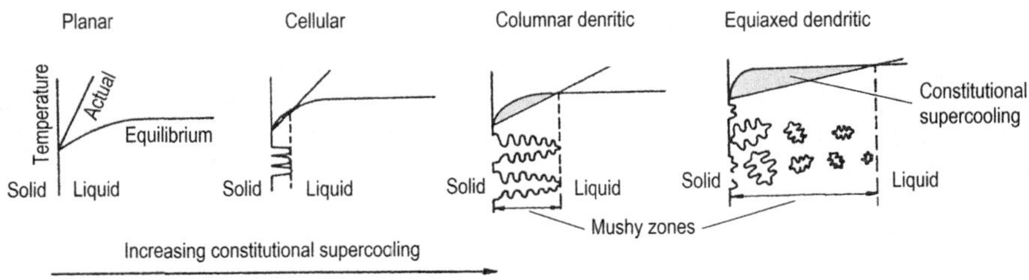

Fig. 1.16 Typical solidification structures generated in alloys as a function of the extent of constitutional supercooling

(in principle homogeneously) in the liquid. The so-formed structure takes the name of equiaxed dendritic since the dendrites do not possess any specific orientation.

1.2 Casting Technologies

The manufacturing process of any metal object involves at a certain step the melting and casting of the material to obtain a solid product for subsequent forming or machining operations, or even with a geometrical shape very close to the final object.

Casting is therefore the process of pouring a molten metal/alloy in a mould, allowing it to solidify into the desired shape. A first class of castings involves the production of simple-geometry shapes such as billets, bars, plates to be considered as semifinished products for further manufacturing stages. In this instance the casting operation falls within the categories of (discontinuous) ingot casting or continuous/ semi-continuous casting. The alternative class, the shape casting or foundry casting, involves the use of moulds generally having more complex shapes, close to the geometry of the final component.

In the following sections these two classes will be described, trying to correlate the technologies with the aspects related to the materials perspective, such as the ability to govern the solidification process, the occurrence of possible defects, the achievable microstructure and properties.

1.2.1 Shape Casting

Shape metal casting processes are efficient in producing a wide range of components, ranging from low-volume, even single part prototypes, to cost-effective mass production by millions of samples. This is made possible by the availability of a broad variety of casting methods, falling into two main groups: the expendable mould- and the permanent mould processes.

Expendable mould processes include sand, plastic, shell, plaster, and investment (also known as lost-wax) molding techniques. These methods make use of temporary, non-reusable moulds that reproduce the negative shape of a solid pattern, including all the accessories needed for the correct feeding of the molten metal during the casting operation. After pouring and solidification of the metal inside the cavity, the mould needs to be broken to extract the casting. As opposed to expendable moulds, permanent moulds are manufactured using a durable material that can withstand the thermal stresses induced by the hot metal cast for a repeated number of cycles. Typically, hot-work tool steels or cast irons are selected for the machining of permanent moulds for relatively low-melting point Aluminium, Zinc, and Copper alloys. The moulds are designed by at least two splitting halves that are kept tightly closed during metal pouring and solidification and can be separated for the extraction of the solid part.

Figure 1.17 recalls the main existing shape casting processes available for metals (Prucha et al., 2008). For each process there are typical thresholds in terms of minimum and maximum size of the castings, achievable surface roughness and geometrical precision, minimum production batch size to make the process convenient, considering the cost of moulds, fixtures and related casting equipment in relation to the performance expected by the product.

Amongst the large number of technologies available, and the even larger existing variants, a selection of the most popular processes has been proposed in Fig. 1.18. Sand mould casting is depicted in Fig. 1.18a. In this process, a re-usable solid pattern replicating the shape of the object to be cast generates a negative print in the compacted sand contained in a two-part moulding box. Cores are used to prevent metal filling in volumes of the cavity which are expected to remain hollow in the

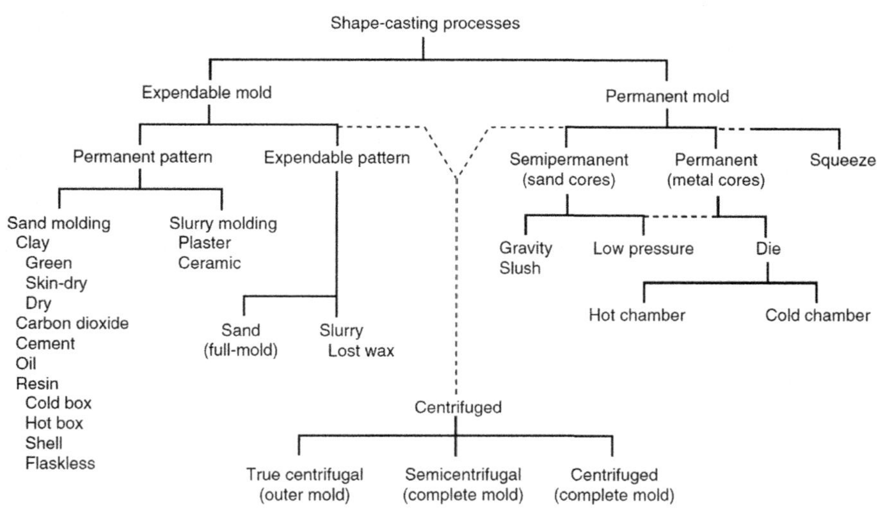

Fig. 1.17 General classification of the shape casting processes

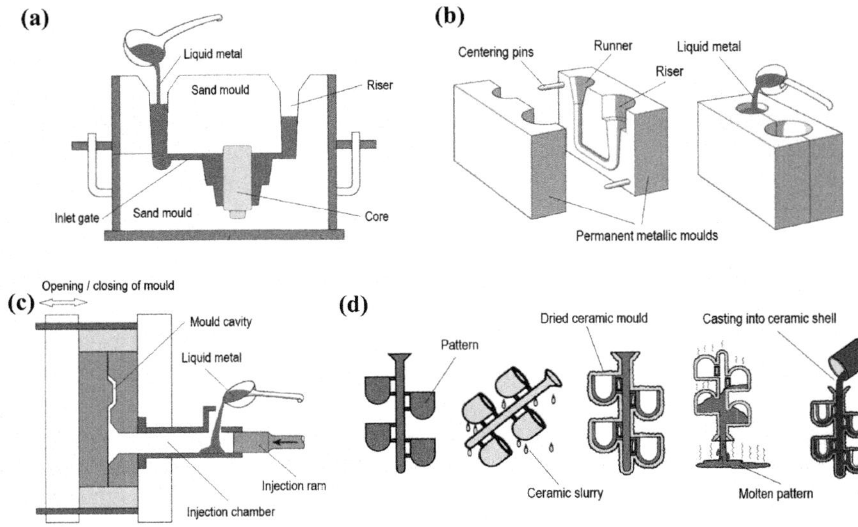

Fig. 1.18 Schematics of several shape casting processes (adapted from The Open University). Sand mould casting (**a**), permanent mould casting (**b**), pressure die casting (**c**), investment casting (**d**)

final casting, whereas runners, risers, gates, feeders and any other fixture needed for the correct delivery of the melt into the cavity are implemented into the pattern. As stated, once the molten metal is poured inside the cavity and solidification is terminated, the mould is destroyed to extract the casting, all the fixtures are separated, and machining of some surfaces might be required to obtain the finished product. Sand casting is particularly suitable for complex shapes to be produced in low quantities and for large-size castings.

Figure 1.18b depicts a permanent mould casting process where the mould is machined out of a block of steel or cast iron to allow multiple casting cycles with the same mould, which is generally operated by robotic systems. Auxiliary fixtures to enable the proper feeding of the melt are required also in this case, with a similar approach as for sand casting. Accurate machining of the moulds reflects into better dimensional tolerances and improved surface roughness of the castings while the larger thermal conductivity of iron-based alloys with respect to sands implies much faster heat extraction rates, resulting in shorter production cycles, reduced cooling times, and superior mechanical properties. Permanent mould processes become cost-effective when the higher investment costs of the moulds can be compensated by larger production batches.

It can be highlighted that the two described processes rely on gravity for the proper filling of the mould cavity by the liquid metal (hence the name of gravity casting processes). Pressure-die casting (Fig. 1.18c) is a further development in which the molten metal is injected into the cavity of a permanent mould by the support of an overpressure. This leads to even faster solidification and cooling rates due to intimate contact of the melt with the mould surface, and the possibility to replicate finer details and to produce thinner casting walls. Pressure-die casting machines are comparatively much more expensive than the gravity casting systems owing to more complex architecture and to the need of large presses to pressurize the melt and keep the dies tightly closed during and right after the liquid metal injection. A further important drawback of this technology lies in the lower microstructural and mechanical quality of the castings as a result of the turbulent flow promoted by the fast injection of the liquid metal into the mould. Further details about the defects induced by the irregular flow will be given in the next section.

Finally, Fig. 1.18d schematically shows the main steps of the investment casting process. Here, patterns are usually formed using wax, polymers or other materials which can be easily shaped. The single pattern or, more frequently, bunches of small-size patterns are assembled into trees, also including the gating system, and dipped into a ceramic refractory slurry which dries around the original model. The pattern is then melted or burnt away, leaving a hollow ceramic mould in which the molten metal is cast. Finally, the resulting bunch of castings is extracted by destroying the ceramic shell and separated by cutting. The investment casting is typically selected for small-size parts having complex shape and for small production of artistic pieces.

1.2.2 Ingot Casting

As stated, semi-finished products are mainly produced by ingot casting or continuous/semi-continuous casting. Small size ingots shown in Fig. 1.19a, b are used as the feedstock for foundries, to be remelted for the production of shape castings. Alternatively, larger ingots, typically in the steel industry, are produced by casting the liquid metal into moulds to form conical blocks, with a weight ranging from hundreds of kilograms up to several tons, as depicted in Fig. 1.19c, d. These ingots are then subjected to forging, shaping large components into their final aspect, or to transformation into oblong products with a rectangular cross-section, to be further hot and cold rolled.

Steel ingots are produced in steelmaking shops by casting a proper volume of liquid steel into a mould, as depicted in Fig. 1.20. Depending on whether the mould is filled from above or below, a distinction is made between top pouring and bottom pouring (Hurtuk, 2008). With top pouring the steel stream is delivered from the ladle directly to the mould by gravity. This method is very rapid and efficient when multiple casting is required, but the quality of the ingots is impaired by oxygen pickup from the environment during pouring and by the induced turbulent flow of the steel upon mould filling. With bottom pouring the mould is filled from the bottom via a runner system. The uprise of the melt from bottom promotes a gentle filling of mould. A protective flux is often used in bottom pouring to prevent oxidation of the upper meniscus surface, to lubricate the flow of liquid against the mould walls, and to trap non-metallic inclusions in the slag, achieving high-quality steel ingots.

Once the mould has been filled with liquid steel, ingot solidification is ruled by the heat removal rate, as affected by the conductivity of the mould, its configuration,

Fig. 1.19 Representative images of cast ingots, to be used as feedstock for Aluminium foundries (**a** and **b**), to be forged into steel parts (**c**) after solidification in a cast iron mould (**d**)

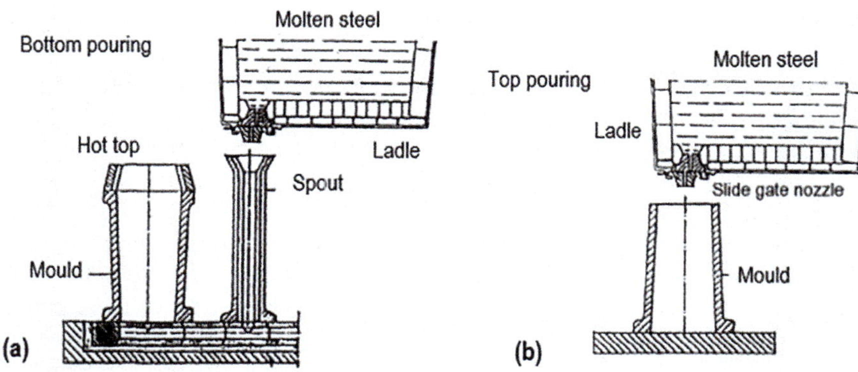

Fig. 1.20 Schematics of steel ingot casting by bottom pouring (**a**) and top pouring (**b**)

and the possible presence of the so-called hot top which is aimed at reducing or compensating the heat losses from the upper surface of the ingot by exothermal reactions (see Fig. 1.20a).

The usual solidification structure found in steel ingots is illustrated in Fig. 1.21. It is generally composed of three distinct zones: a chill zone, a columnar dendritic zone and a central equiaxed zone. The chill zone is a thin layer of metal that almost immediately forms as the liquid steel contacts the relatively cold walls of the mould. Recalling the effects of the thermal gradient in the liquid discussed at Sect. 1.3, the steep gradient locally built in this surface layer promotes planar solidification by the nucleation of numerous small grains. By increasing the thickness of the solid steel in the ingot, the thermal gradient is progressively reduced. This is due to the contribution of the latent heat released on solidification, to the slower heat removal rate towards the mould walls due to the decreased thermal diffusivity of the thicker solid steel layer, and to the air gap formed at the mould-ingot interface caused by shrinkage. The decreased gradient, together with the increased segregation of elements into the liquid ahead of the solidification front favour the transition toward a columnar dendritic structure (see Figs. 1.16 and 1.22a) with grains larger in size and an elongated shape aligned to the heat flow direction, namely normal to the mould walls.

Eventually, as the solidification front converges at billet centre, the liquid regions affected by constitutional supercooling tend to overlap, as schematically depicted in Fig. 1.22b. As a result of the extended supercooling so created, conditions for the formation of equiaxed grains can be met at ingot center.

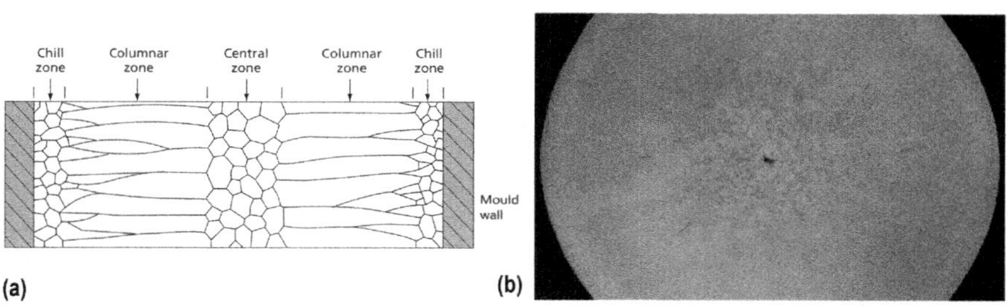

Fig. 1.21 The different microstructural zones found in a steel ingot. Schematic of a longitudinal section (**a**) macrograph of a transversally sectioned steel billet (**b**)

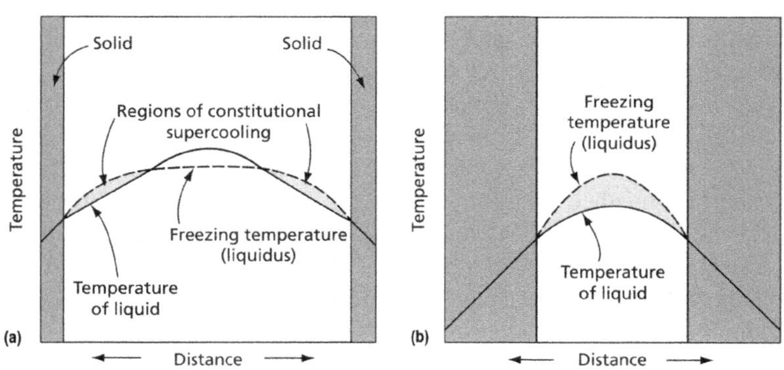

Fig. 1.22 Schematics of a typical steel ingot grain structure (**a**) and image of the cross section of a cast steel ingot (**b**)

Extensive columnar structure in billets is seen as a negative feature since it promotes higher degree of segregation and it is more prone to solidification cracking. Minimizing its formation is therefore an important research topic which implies the control of the steel chemistry, of the pouring temperature, of the mould characteristics influencing the heat dissipation, hence the steel cooling rate.

1.2.3 Continuous and Semi-Continuous Castings

While ingot casting is the preferred method for certain special alloy steels, tool steels, forging alloys and for remelted ferrous and non-ferrous alloys, for the remaining part of the worldwide production of semi-finished products, accounting for the vast majority of steel produtcion and a large share of aluminium alloys, the preferred manufacturing route involves the continuous or semi-continuous casting.

In the continuous casting process, the molten metal is transferred from the ladle into a tundish which acts as a reservoir for the underlying casting machine (Fig. 1.23a). Downstream to the tundish there is a mould cavity (or multiple moulds in case of multi-strand installations) where the first stages of solidification take place. The mould is an open-ended, water cooled copper cavity that promotes fast heat extraction and rapid solidification of a first external shell of the steel. The solid shell, that still contains a large volume of the molten metal, is continuously withdrawn from bottom of the mould at a constant rate that matches the upstream flow rate of the liquid metal. To prevent the sticking of the freshly solidified shell to the mould walls, this latter is subjected to a vertical oscillatory motion and a flux powder is added on top of the mould to protect from reoxidation and act as a lubricant once melted. The length of the mould and the casting speed need to be accurately tuned to enable the build of a solid shell thickness that can withstand the internal pressure of the molten metal once it exits the mould. To start a casting operation, a dummy bar is initially inserted into the mould so that the first amount of molten steel is confined in a closed volume and can be safely extracted downwards through a chain line until a steady condition is attained. As such, once started, the process ideally operates in a steady state mode, ensuring high productivity, uniform quality of the bars and significant energy savings.

Below the mould, a set of nozzles deliver water sprays to the surface of the strand (generating the so-called secondary cooling) to complete the solidification at the core of the strand within a length which is termed the metallurgical length. The metallurgical length can exceed a dozen of meters for industrial steel plants, but it is limited to fractions of a meter for more thermally conductive metals such as aluminium- or copper-based alloys.

Along the secondary cooling length, the strand is supported by rolls which drive it along the designed path, eventually reaching a cut-off torch or shear unit, where the continuous strand is sectioned to the desired length of the bars or slabs. Different plant arrangements have been developed over the years and are still in use for continuous

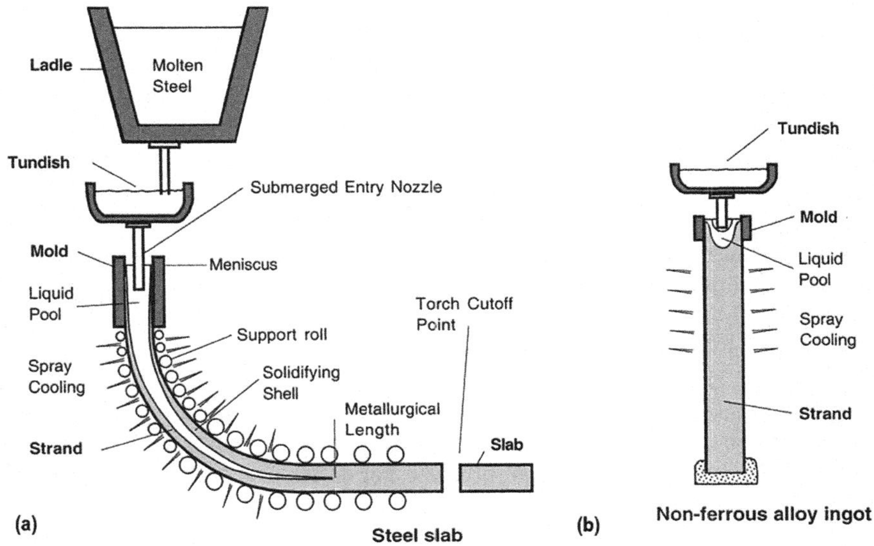

Fig. 1.23 General layout of a continuous casting plant for steel slabs (**a**) and of a semi-continuous plant for aluminium alloy ingots (**b**)

casting of steels, including vertical, curved (as shown in Fig. 1.23a), and horizontal configurations, which are mainly preferred for non-ferrous alloys.

When it comes to aluminium and other nonferrous alloys, semi-continuous casting, also known as direct chill casting, is the reference process for billet production. Figure 1.23b highlights the main differences compared to continuous casting. Solidification begins in a water-cooled mould of reduced length (often resembling just a ring) with respect to steel casting. In vertical direct chill casting, the process starts with a dummy bottom block which initially confines the molten alloy and is withdrawn by hydraulic rams at a constant rate. The bottom block is lowered into a pit until the desired billet length is reached, then the molten metal flow is stopped, the mould and feeding assembly are tilted or slid away, and the solidified billet is raised out of the casting pit. In direct chill casting, multi-strand configurations are the rule rather than the exception. Issues related to sticking of the solid shell to the mould are faced by the selection of proper lubricants and secondary cooling is carried out by water flows and sprays, depending on specific design.

Finally, it is to recall that horizontal variants of the direct chill casting are also quite common. In this configuration the length of the billet is not bound anymore to the depth of the casting pit and longer products can be readily cast, drawn by a conveyor system and cut into billets of the desired length by online systems.

1.3 Defects in Castings

It must be considered that different metals and alloys feature distinct "castability", a property related to the ability to be successfully cast into the desired shape, which is affected by their behaviour in the liquid phase and during the following solidification and cooling stages. In particular, the main physical properties governing castability are the fluidity of the liquid melt, the heat capacity and thermal conductivity of liquid and solid phases, the solidification temperature range, the coefficient of thermal contraction of the solid and the possible phase transitions met upon cooling to room temperature.

It can be preliminarily understood that the fluidity of the melt affects the ability to fill the tiny gaps and thin walls of the casting mould. The extent of solid–liquid contraction and the solid-state shrinkage rule the development of shrinkage voids and the build-up of residual stresses at the solid stage. The precipitation of low-melting phases and brittle constituents can promote hot cracks and other forms of brittle behaviour upon cooling of the castings. Moreover, several defects are generated by a combination of the above-described effects, making the rigorous classification of the casting defects extremely complicated. An exhaustive compendium of the existing casting defects is out of the scope of the present text (see for instance: Campbell, 2003; Gariboldi et al., 2010). However, several solidification-related phenomena which occasionally can results in important defects will be here recalled to enrich the discussion. In particular, we can first refer to the group of defects directly connected to solidification, including gas porosity, shrinkage porosity and hot tearing/cracking. A further group of defects of great practical implication in casting is the entrainment of exogenous matters and oxide films in the solid, which leads to the generation of inclusions.

1.3.1 Gas Porosity

One of the main causes of gas porosity stems from the precipitation of the gases initially dissolved in the liquid, once the temperature decreases. Associated to the transition to the solid phase, there is generally a sharp drop of solubility of atmospheric gases (i.e. nitrogen, oxygen or hydrogen) in metals, as shown in Fig. 1.24 for hydrogen in iron and aluminium.

Hydrogen is considered as one of the most dangerous gases found in metals. It can be absorbed in the liquid melt by different sources such as humidity on feedstock ingots, damp refractories in furnaces or ladles, burning of hydrocarbon fuels, reaction with wet moulds and core materials during casting. An additional source of gas entrapment in the liquid metal can be the turbulent flow generated during the casting process. This occurs when the molten metal is poured too quickly into the mould, leading to extensive trapping of air.

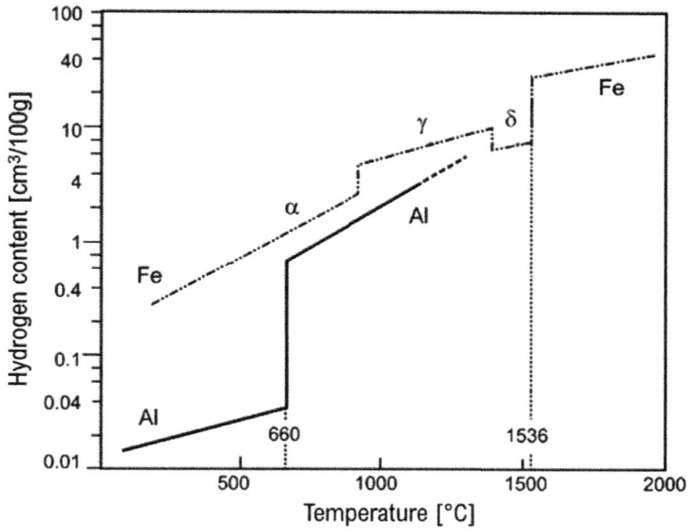

Fig. 1.24 The solubility of hydrogen versus temperature in iron and aluminium at ambient pressure

When solidification of the gas-containing melt starts, gas bubbles are rejected ahead of the liquid front since the solid metal cannot hold as much gas as the liquid counterpart. These bubbles spontaneously tend to float upwards and escape through the liquid/slag interface to the open environment. However, when small bubbles become trapped within the dendrite arms, the metal will fully freeze around them, embedding these gas pockets in the casting structure. Figure 1.25a schematically depicts the described mechanism, while Fig. 1.25b shows the typical aspect of a gas porosity. The trapped bubbles appear on a section of the casting as rounded, circular cavities or holes, ranging in size from few micrometres to several millimetres.

Gas porosity can be eliminated through accurate casting and mould design and by suitable degassing treatments of the molten alloy before pouring into the mould. A popular method to evaluate the residual gas content, especially used in Al foundries

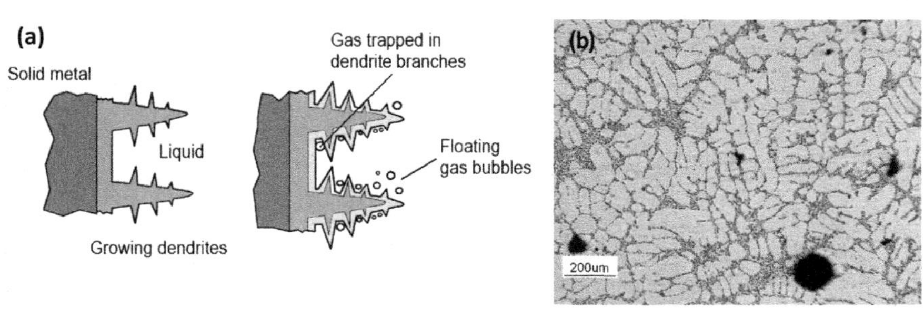

Fig. 1.25 Schematic mechanism of the interdendritic trapping of gas bubbles (**a**) and aspect of a gas porosity in an Al alloy casting sample (**b**)

Fig. 1.26 Sectioned views of two reduced pressure test samples, sample in (**a**) was poured after proper degassing of the melt, while in (**b**) significant amount of residual gas was still present before pouring in the cup

to check the hydrogen content, is the "reduced pressure solidification test", whereby a small volume of the liquid is taken from the melt, immediately poured in a cup and placed under a vacuum dome. The sample is then allowed to solidify under low pressure (around 900 mbar) so that the volume of hydrogen gas expands approximately 10 times more than under solidification at atmospheric pressure. This enhanced expansion can easily become a measure of the porosity potential of the melt, once cast. Figure 1.26 provides comparative views of two Al alloy casting samples. In the first section (Fig. 1.26a), limited porosity is observed only in subsurface regions on top of the casting, and the upper profile reveals the expected concavity accounting for the liquid shrinkage. The second sample (Fig. 1.26b) refers to a poorly degassed alloy. Here, extensive gas macro-porosity can be observed, and the convex upper profile suggests that gas expansion was more than counterbalancing the melt shrinkage upon solidification.

Finally, it might be curious to recall that the same physical concepts standing behind gas porosity in metal castings can also be observed using a much more popular carbonated brown drink, which is well known to be saturated with CO_2 when at the liquid stage. Figure 1.27 illustrates what happens when experimenting the solidification of this particular "melt" in a domestic freezer under different starting conditions. The figures confirm that also for this case, degassing can dramatically limit the development of gas porosity upon solidification.

1.3.2 Shrinkage Porosity

A close volume of liquid subjected to solidification and cooling undergoes three different types of contractions before reaching the room temperature, as schematically shown in Fig. 1.28. A first contraction is observed in the fully liquid stage, as the overheated melt cools reaching the liquidus temperature. In practical castings,

Fig. 1.27 Evidence of different gas solubility in a common carbonated drink: in liquid form (**a**), the same initial amount of drink frozen right after opening of the bottle (**b**), and after degassing the drink (**c**)

this contraction does not provide any specific concern since the extra liquid needed to compensate for the small volume reduction, can easily flow from the feeder.

Of much greater importance is the second type of contraction, occurring within the alloy solidification range. The more compacted arrangement of atoms in the solid crystal compared to the liquid, generally leads to a quite extensive shrinkage. Table 1.2 summarizes typical data of the solidification shrinkage for some metals and alloys. It can be inferred that in general, those materials that crystallize by a close packed structure (i.e. face centred cubic and hexagonal close packed) show the largest values of the solidification shrinkage. Alloying elements also play a paramount role, affecting the solidification interval and the overall shrinkage within this range. Finally, it is noted that grey (graphitic) cast irons feature significantly low values of shrinkage compared to steels due to the expansion effect promoted by the precipitation of graphite which can partially or even totally compensate for the matrix solidification shrinkage.

Fig. 1.28 The three types of contractions contributing to shrinkage of castings (adapted from Gowsalya and Afshan, 2020)

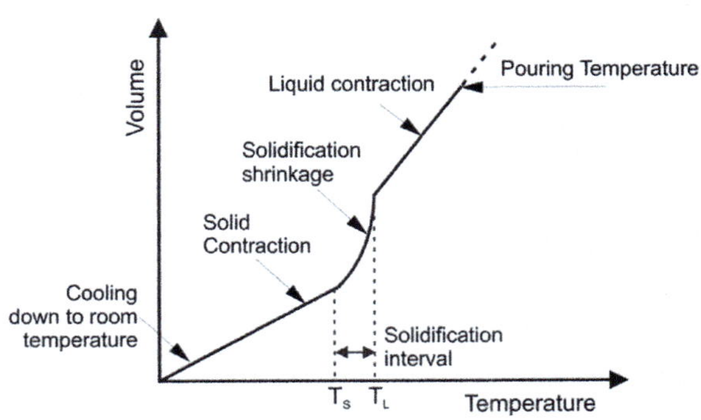

Table 1.2 Solidification shrinkage and solid contraction for some metals and alloys

Metal/alloy	Melting temperature T_m (°C)	Crystal structure at room temperature	Solidification shrinkage (%)	Solid contraction (%)
Al	660	FCC	8.0	1.5
Al-12Si	580–570	FCC	3.5	1.1
Al-4Cu	650–550	FCC	8.8	1.2
Cu	1083	FCC	4.0	1.7
Cu-35Zn	900–885	FCC	6.5	1.7
Ni	1453	FCC	5.1	1.9
Fe	1536	BCC	3.2	1.8
C45 steel	1460	BCC	6.0	1.7
Grey cast iron	1250	BCC	1.8	0.8
Mg	651	HCP	4.10	1.6
Zn	420	HCP	4.08	1.3

Right after completion of the solidification stage, the alloy still needs to cool down to room temperature. This solid-state cooling implies a third type of contraction which depends on the coefficient of thermal expansion of the material and on the width of the range from solidus to ambient temperature. The amount of the solid contraction for a number of metals and alloys is also provided in Table 1.2. As the casting attempts to reduce its size, geometrical constraints can be present due to the mould or the shape of the casting itself. As a result, stresses can build up within the casting, making it prone to cracking, to warpage once freed from the mould, and more susceptible to failures during service due to the effects of such residual stresses.

The analysis of the data presented in Table 1.2 suggests that solidification shrinkage is the contraction step of largest concern for the integrity of castings. In order to compensate for the wide shrinkage, feeding systems are required to provide extra material volumes during solidification.

When the feeding is not appropriately designed, shrinkage macro-voids and porosity defects can grow in castings, producing macro- and micro-porosity, respectively. Shrinkage micro-pores are cavities that are produced due to the lack of liquid available to locally fill the empty volumes generated due to solidification contraction. As such, they are generally placed in the last regions that solidify, namely in the interdendritic regions of the microstructure, as shown in Fig. 1.29. If the shrinkage porosity is small in size and confined to the thickest sections of castings, it usually does not cause any specific concern. Indeed, if homogeneously distributed and in small amount, shrinkage micro-porosity is generally accepted and considered as an intrinsic feature of castings. However, if pores are larger in size and interconnected, they can severely deplete the strength and become a serious concern, especially for castings which need to be gas or water tight.

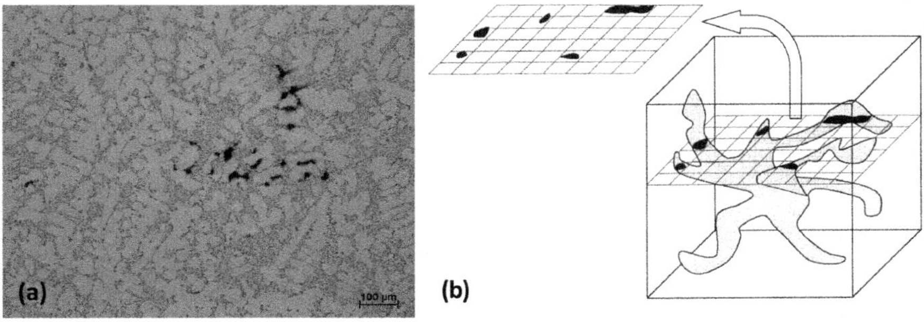

Fig. 1.29 Shrinkage micro-porosity observed in Al alloys (**a**) and schematic image of 3D interconnected shrinkage pores (adapted from Anson & Gruzleski, 1999) (**b**),

Considering now the issue about feeding, Campbell proposed a list of five possible feeding mechanisms which can be active to compensate for metal shrinkage during solidification and cooling, as summarized in Fig. 1.30 (Campbell, 2003). To start with, it must be clear that since extra liquid metal needs to be fed to the solidifying volume, it is the rule to design casting with one or more separate reservoirs of hot melt, the feeders (also called risers). Liquid feeding through the feeders is therefore the primary mechanism to counteract shrinkage. As solidification progresses, the so-called mass feeding becomes predominant, whereby the movement of a semi-solid slurry composed of solid crystal aggregates and residual liquid is set. Mass feeding is mainly observed in thick sections of castings and tends to stop when the volume fraction of solid reaches around 50%. Once the solid crystals/dendrites eventually start creating a rigid network across the casting section, mass feeding becomes increasingly difficult and the next regime identified as interdendritic feeding becomes predominant. In this condition, the only gaps left for the feeding of extra material to the shrinking regions are the interdendritic channels. Further cooling leads to the increase in liquid viscosity and reduction in size of the channels so that the material flow becomes unfeasible. At this point, the high hydrostatic stress formed at the last stages of solidification can be relaxed either by formation of shrinkage pores or by the inward plastic flow of already solidified walls of the casting, generating the mechanism of solid feeding. For sake of completeness, burst feeding must also be listed in parallel to mass feeding. In burst feeding it is assumed that when the hydrostatic tension exceeds a critical value in poorly fed regions, some of the weakest barriers might suddenly collapse, allowing new metal to flood into the poorly cavities.

1.3.3 Hot Tearing

Hot tearing, also known as hot cracking or hot shortness, denotes the formation of intergranular cracks in castings in the last stages of solidification. It is recognized that

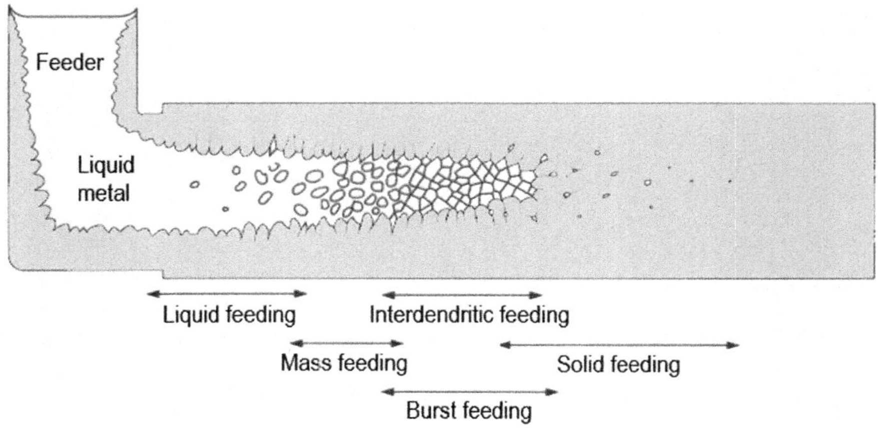

Fig. 1.30 Schematic of the different feeding mechanisms found in metal castings, (adapted from Campbell, 2003)

hot tears nucleate above the solidus temperature of alloys, when the volume fraction of solid is above around 95% and the solid phase is already organized in a continuous network of grains. Figure 1.31 shows the schematic steps about the progression of solidification and the development of a hot tear. Only after the temperature has dropped below the coherency temperature (when the continuous network of grains is set, shown in Fig. 1.31b), the solidification shrinkage and the thermal contraction can impose strains and stresses on the solid alloy, which are the requisites for hot tearing.

Further conditions to be evaluated for the hot cracking susceptibility of casting alloys are the extension of the freezing range, the size and shape of grains and the amount of liquid available in the last stages of solidification. It is recognized that a large freezing range of an alloy favours hot tearing since the alloy will spend a longer time in the vulnerable state, where a thin liquid film exists between the dendrites. It is also known that a fine grain structure can help to avoid hot cracking due to the ability of the small volumes of liquid to distribute over the larger grain-boundary surface area and due to the increased ability of fine grains to accommodate the shrinkage strains by slip along their boundaries.

The amount of liquid available in the last stages of freezing plays a paramount role on hot tearing susceptibility. Alloys with significant amount of eutectic liquid in the final stages of solidification are usually free from hot tearing concerns since the presence of a continuous network of solid crystals coexisting with liquid pockets (condition b in Fig. 1.31) is virtually suppressed. Other theories suggest that the presence of large volumes of residual liquid could even backfill occasional cracks which developed upon solidification, thus promoting an additional crack-healing action.

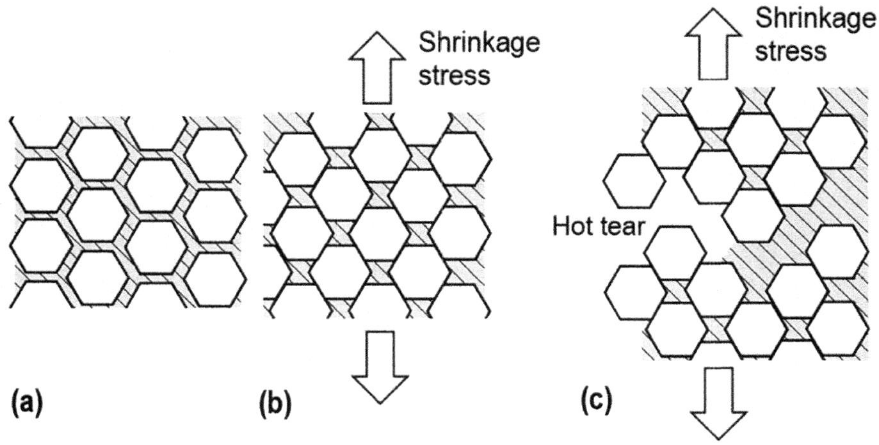

Fig. 1.31 Schematic stages about the progression of solidification and the development of a hot tear. Initially, solid grains are fully surrounded by liquid and no shrinkage stress can develop (**a**), then, a continuous network among solid grans is built and shrinkage stresses are imposed on the solid fraction (**b**), eventually, hot tears can nucleate and grow along the weak grain boundaries and through the liquid pockets (adapted from Campbell, 2003)

1.3.4 Inclusions

For the development of hot tears and any other type of crack which might lead to catastrophic failure of the casting under load, nucleation sites are always required. The source of nucleation is generally associated to a heterogeneity in the alloy microstructure ranging from the macroscopic size scale, such as a geometrical notch on the surface or a macroscopic porosity in the volume acting as a stress raiser, down to the nanometric size where any residual pocket of liquid or a tiny second-phase particle can fail under load or debond from the matrix, initiating a crack. In castings, the above-described gas and shrinkage porosity can often act as nucleation sites. Additionally, inclusions can represent a common source of failure.

Inclusions can be classified in endogenous and exogenous. Endogenous inclusions are formed "in situ", namely within the liquid metal by chemical reaction with dissolved gases or with specific elements used for the liquid metal treatment. Examples of endogenous inclusions are provided by the deoxidation and desulphurisaton reactions promoted in steelmaking practice. To deoxidize the molten steel, titanium, aluminum, and silicon are added to the melt to capture oxygen and promote its precipitation as Al_2O_3 or SiO_2 inclusions. In principle, these light phases tend to float to the slag covering the liquid steel in the ladle, but the smallest aggregates featuring slower floating rates remain trapped into the growing solid front, resulting as inclusions in the final microstructure. During solidification, further inclusions can precipitate mainly by virtue of the desulphurization treatment. A carefully designed steel composition can stimulate the precipitation of manganese and calcium sulphides

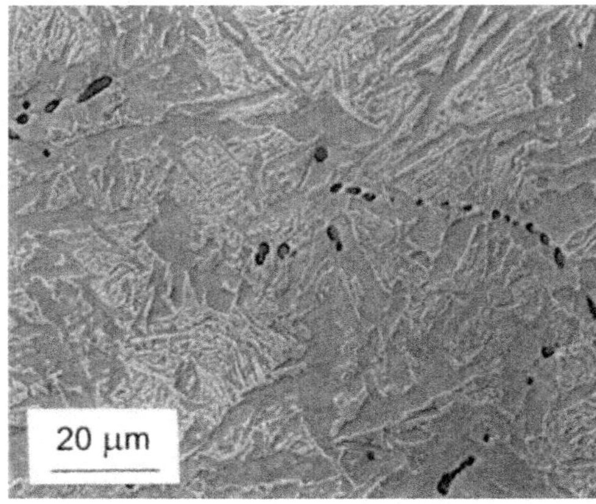

Fig. 1.32 Microstructure of a C–Mn steel structure after solidification in continuously cast billets with evidence of MnS inclusions as the dark phases

(MnS, CaS), independently or in combination with alumina and silica, as shown in Fig. 1.32.

Exogenous inclusions include all those particles that already existed as a separate phase before the onset of solidification, deriving from external sources, hence "ex situ". Examples of exogenous inclusions can be small pieces of eroded refractory lining or portions of slag entrained into the casting during pouring.

An additional and typical source of exogenous inclusions found in some reactive metals such as aluminium is generated by the presence of a native oxide film on the surface of the melt once it is exposed to the air environment. It is worth recalling that a natural surface film is not harmful while it stays on the surface of the liquid metal since it can protect the rest of the melt from further extensive oxidation. However, during handling of the liquid, especially during casting, the liquid surface might be subjected to an irregular flow, creating folds which promote the entrainment of the original surface oxide into the bulk liquid, as schematically depicted in Fig. 1.33.

The presence of the oxide film into the casting volume introduces new weak interfaces since the oxide layer cannot be dissolved nor be firmly bond to the metallic matrix once it reaches the solid state. In addition, the entrainment process should be considered as a folding action producing two overlapped oxide surfaces (see inset of Fig. 1.33), which are effectively identified by the term "bifilm". Due to the substantially negligible bond between the two oxide surfaces, the oxide bifilm is considered as an existing crack leading to significant reduction of the casting strength and ductility.

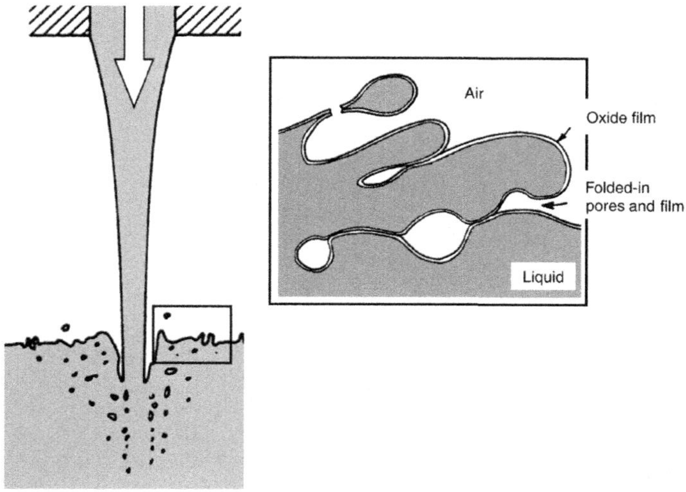

Fig. 1.33 The simplified mechanism of surface oxide entrainment formed for example during the turbulent pouring of a liquid metal (adapted from Campbell, 2008)

References

Abbaschian, R., Abbaschian, L., & Reed-Hill, R. (2009). *Physical metallurgy principles* (4th ed.). Cengage Learning Publisher.

Anson, J. P., & Gruzleski, J. E. (1999). The quantitative discrimination between shrinkage and gas microporosity in cast aluminum alloys using spatial data analysis. *Materials Characterization, 43*, 319–335.

Ashcroft, N. W. (1969). Liquid metals. *Scientific American, 221*, 72–85.

Campbell, J. (2003). *Castings* (2nd ed.). Buttherwort- Heinemann Publisher.

Campbell, J. (2008). *ASM handbook—casting practice—guidelines for effective production of reliable castings* (Vol. 15). ASM International.

Chalmers, B. (1954). Melting and freezing. *Journal of Metals, 6*, 519–532.

Gariboldi, E., Bonollo, F., & Parona, P. (2010). *Handbook of defects in high-pressure diecastings*. Associazione Italiana di Metallurgia Publisher.

Gowsalya, L. A., & Afshan, M. E. (2020). Heat transfer studies on solidification of casting process. In N. A. Z. Abdallah, *Casting processes and modelling of metallic materials*. IntechOpen. https://doi.org/10.5772/intechopen.91879

Hurtuk, D. J. (2008). Steel ingot casting. In *ASM handbook—casting* (Vol. 15). ASM International Publisher.

Laughlin, D., & Hono, K. (2014). *Physical metallurgy* (5th ed., Vol. 1). Elsevier.

Messler, R. W. (1999). *Principles of welding: processes, physics, chemistry, and metallurgy*. Wiley Publisher.

Prucha, T. E., Twarog, D., Monroe, R. W. (2008). History and trends of metal casting. In *ASM handbook—casting* (Vol. 15). ASM International Publisher.

Strickland, J., Nanchev, B., & Dong, H. (2020). On directional dendritic growth and primary spacing—A review. *Crystals, 10*, 1–30. https://doi.org/10.3390/cryst10070627

Chapter 2
Plastic Deformation of Metals

2.1 General Concepts of Plastic Deformation Processing

In a previous chapter dealing with solidification, it has been shown that metallic parts with a shape substantially identical to that required for the final use of the designed component can be produced by foundry castings. An alternative route for the manufacturing of metal parts involves a large group of processes based on plastic deformation. The raw materials for this route are billets and ingots coming from the solidification processes (see Sects. 1.2 and 1.3) which are now subjected to extensive plastic deformation steps under different conditions to obtain new semi-products, also identified as wrought products, like bars, wires, plates, sheets, extruded profiles rather than almost finished objects by forging, sheet forming, deep drawing.

Even if there exist hundreds of plastic deformation processes with an even higher number of variants, it is instructive to introduce a first broad classification based on the type of forces exerted on the workpiece while it is formed. As summarized in Fig. 2.1, we can identify five groups of processes (Dieter, 1988).

- Direct-compression processes
- Indirect-compression processes
- Tension-based processes
- Bending-based processes
- Shearing-based processes.

In direct-compression processes the force producing the plastic deformation is directly applied by the tools to the surface of the workpiece and the metal is forced to flow at an almost right angle to the direction of compression (e.g. in forging and in rolling, Fig. 2.1a, b). Indirect- compression refer to those processes where the primary forces are of different types, even tensile, but the reaction of the tools induces a net compressive action on the workpiece, typical examples are extrusion and wire drawing (Fig. 2.1c, d).

M. Vedani, *Metal Science in Modern Manufacturing Technologies*, Springer Series in Materials Science 351, https://doi.org/10.1007/978-3-031-97125-9_2

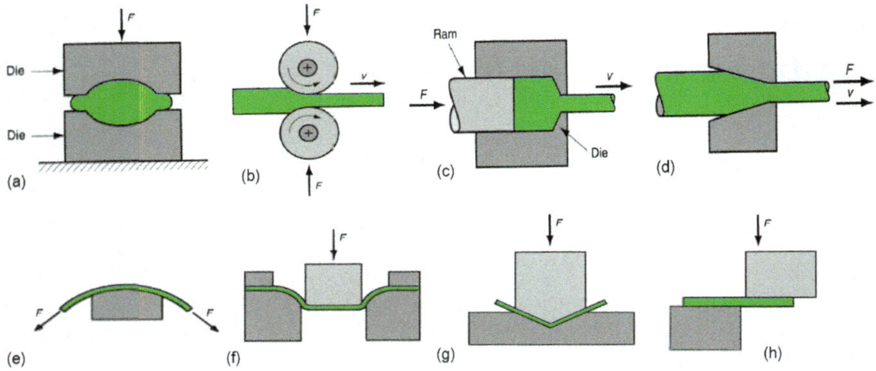

Fig. 2.1 Schematic view of the main forming technologies. **a** forging, **b** rolling, **c** extrusion, **d** drawing, **e** stretch forming, **f** deep drawing, **g** bending, **h** shearing/cutting (adapted from Dieter, 1988)

Tension-type forming is used in stretching or in deep drawing of sheets when mainly tension strains are used to form a metal blank into the desired shape (Fig. 2.1e, f). Bending is a further method exploited to form blanks into simple shapes (Fig. 2.1g), whereas shearing forces are frequently used to cut them by fracture in the plane of shear (Fig. 2.1h).

Considering a different perspective, we could look at the plastic forming processes not only as a way to modify the shape of workpieces, but also to change their original microstructure and properties as it is inherited from the solidification stage. Indeed, there exist a first group of plastic deformation processes identified as primary or rough processes, carried out at high temperature, which are aimed at reducing the ingot solidification structure to a more refined and homogeneous condition, in the form of simple-shaped wrought semi-products. Further processing, known as secondary or finish forming, might be applied to those workpieces that need improved dimensional tolerances, surface finishing and enhanced properties. Secondary forming is usually performed at room temperature, representative technologies are wire drawing, cold rolling, and sheet forming.

It must be realized that according to the different processing conditions, as affected for instance by temperature, amount of applied strain and strain rate, the metal workpiece undergoes a broad range of possible transformations in terms of its grain orientation, shape and size of microstructural constituents. These change in turn rule its further processability, the occurrence of manufacturing defects and the resulting physical and mechanical properties of the designed objects. In the following sections such transformations will be described and discussed with a particular attention to the opportunities of manipulating the alloy properties during plastic deformation processing.

2.2 Cold Plastic Deformation

2.2.1 Concepts About Strain Hardening and Annealing

Strain hardening, also called work-hardening, is the strengthening effect produced when a metal is strained beyond its yield point. As plastic strain is accumulated, an increasing stress is required to produce additional deformation and the metal apparently becomes stronger and more difficult to deform. In practical terms, it is possible to measure the extent of strain hardening in a metallic alloy by testing its properties using samples collected at different stages of a plastic deformation process (e.g. at various levels of reduction of thickness in cold rolling, or at specific amounts of reduction of area in cold drawing of a wire). A typical trend of the evolution of tensile properties and hardness for an aluminium alloy is depicted in Fig. 2.2. It is readily observed that strength and hardness monotonically increase with the amount of cold work, while ductility is subjected to a significant drop, down to very low values of elongation to fracture, even for a ductile alloy. In the context of a real cold-forming process, any metallic alloy will show a limit in the amount of cold deformation, above which any further processing becomes unpracticable owing to material brittleness and excessive hardness. Therefore, one would require restoring the original soft state of the alloy through a suitable thermal treatment called annealing.

Strain hardening is generally attributed to the multiplication of dislocations in a cold deformed metal and to the effects of their mutual interactions and their accumulation against other microstructural barriers such as the grain boundaries and second phase particles. More precisely, we can state that during plastic deformation the structure of a metal changes in several ways (Humphreys & Hatherly, 2004).

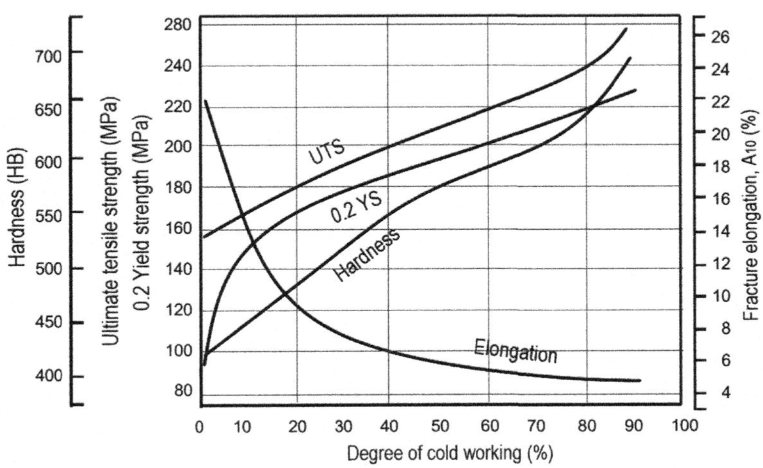

Fig. 2.2 Evolution of tensile properties and hardness of an Al-Mg-S alloy as a function of the degree of cold-working by rolling (data from Siegert et al., 2000)

The first obvious evidence of the induced deformation consists in the change of the grain shape. The grains become elongated along the main plastic-flow direction with an associated large increase in their grain boundary area. A second change is directly related to the increased dislocation density created during plastic deformation. It has been evaluated that an annealed metal contains around 10^4–10^6 dislocations per mm^2, while in a heavily work-hardened metal the density rises up to 10^{10} dislocations per mm^2. This excess of dislocation tends to rearrange into a substructure inside the grains, formed by an almost regular array of high-dislocation density regions surrounding dislocation-lean areas, thus creating a sort of cellular substructure where the dislocation tangles represent the cell boundaries.

Finally, a third effect of plastic deformation relates to the re-orientation of the grain crystals ruled by the relative direction of the applied stresses and strains. The induced rotations of the individual grains statistically promote a preferred crystal orientation (also called crystal texture) in deformed polycrystalline metals, which becomes sharper as the deformation proceeds.

It is worth recalling now that the energy provided as mechanical work through the tools that induce the plastic deformation is mostly converted into heat. This is also why a severely cold-worked workpiece can become very hot upon deformation, even without any deliberate preheating treatment. As opposed to the dissipated heat, only a small percentage of the input energy is stored in the deformed metal as internal energy associated to excess of dislocations and the newly formed interfaces (i.e. grain boundaries). It is this stored energy that enables the activation of all the structural changes observed during the annealing of a deformed metal.

Being the cold-worked metal in a state of higher internal energy, it can be assumed to be in a "mechanically" stable condition but not thermodynamically. Indeed, it would tend to evolve to a lower-energy state close to that of the undeformed condition if only a sufficient amount of thermal activation energy is provided. As already mentioned, the thermal process by which a deformed metal softens and reverts to a strain-free, hence easily processable, condition is called annealing. Annealing consists of an isothermal heat treatment where the cold-worked workpiece is stored for few hours at a temperature between 1/3 and 1/2 of its absolute melting temperature.

Figure 2.3 summarizes the changes observed during the annealing of a deformed metal. Annealing is generally divided into the three distinct steps of recovery, recrystallization and grain growth. The plot shows that recovery occurs at the lowest temperatures or at the initial stages of an isothermal treatment. During recovery, no significant changes are observed in mechanical properties while internal stresses and other physical properties like electrical resistivity are restored almost to their initial state. It is only with the onset of recrystallization that the cold-worked metal will show a clear softening (drop in strength and hardness) and an increase in ductility. The last stage of grain growth represents the natural evolution for a polycrystalline structure held at high temperature. The grains will try to grow larger in an attempt to decrease the surface energy of their boundaries, thus reaching an even more thermodynamically stable condition. It should be clear that this latter step has preferably to be avoided in real annealing treatments, since in general no additional advantages are achieved.

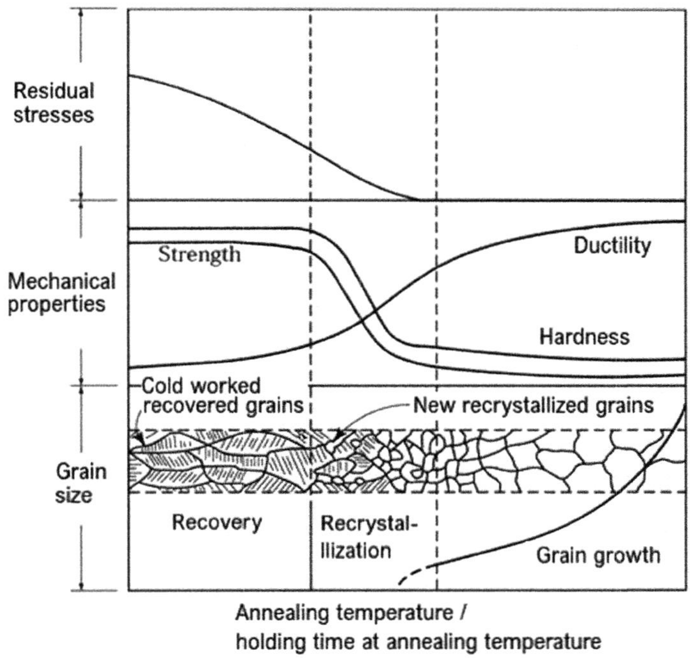

Fig. 2.3 Development of recovery, recrystallization and grain growth during annealing of a cold-worked metal (adapted from Hu, 1963)

A more precise picture of the microstructural evolution of a cold-deformed and annealed metal is given in Fig. 2.4. The first image 4.a, and the corresponding enlarged view a1 recall the initial deformed state of the material, with the evidence of the excess of dislocations.

Figure 2.4b refers to the recovery stage, and the enlarged views from b1 to b3 suggest that while the general grain structure remains unchanged, the dislocation substructure will evolve. In particular, frames b4 to b6 suggest the two main mechanisms taking place during recovery, namely the annihilation and the rearrangement of dislocations. Annihilation is due to the interaction of the stress fields associated to each dislocation, whereby dislocations of opposite sign which may glide on the same plane may annihilate by attracting and approaching each other. Similarly, a combination of glide and climb could be activated by virtue of thermal processes to assist annihilation even for dislocation pairs laying on different planes. Once annihilation has progressed, the remaining dislocations can still lower their associated strain energy by rearranging into a regular array of boundaries, producing a so-called network of low-angle grain boundaries (LAGBs), named after the small misalignment created by the stacking of edge dislocations on specific planes (see image in frame b6). For sake of completeness, we shall specify that, opposed to the LAGBs, there exist the high-angle grain boundaries (HAGBs) generally separating the grain

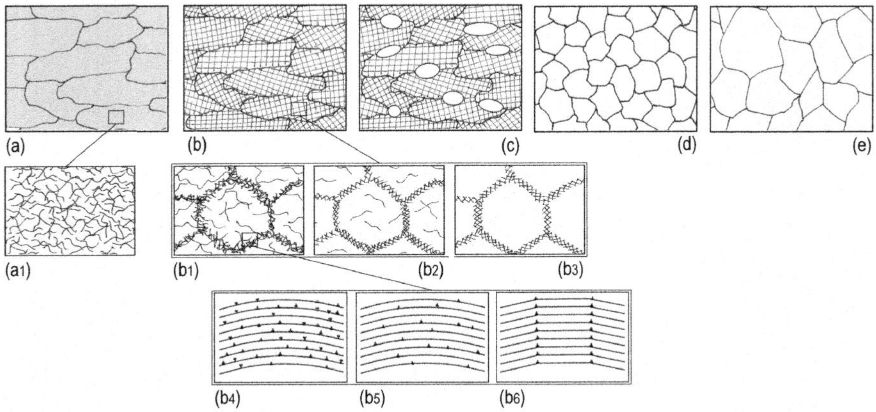

Fig. 2.4 Schematic views of the microstructural evolution upon cold-working and annealing of a metal observed at different size scales. Cold worked condition (**a**), recovery (**b**), recrystallization (**c** and **d**), grain growth (**e**)

crystals by a much larger angle (see the boundaries identified by the thick lines in Figs. 2.4b).

Recrystallization generally follows the recovery stage, even with a certain degree of overlapping, as depicted in Fig. 2.4c, d. Recrystallization can be defined as the phenomenon of replacement of the deformed structure by a new set of strain-free grains and it is accompanied by a distinct drop in hardness and strength of the cold-worked metal, along with a rise in ductility. It usually develops through a discontinuous process, namely by the nucleation of new crystals featuring a sufficiently high misorientation with respect to the surrounding lattice, at distinct heterogeneities of the microstructure such as at the grain boundaries, the second phase particles or by the abnormal growth of specific sub-grains. Once nucleation is activated by the thermal energy provided by the annealing treatment, the strain-free nuclei thus formed will grow by sweeping the deformed structure and leaving behind a new stress-free lattice until the whole volume is replaced by the new grains (Fig. 2.4d). Prolonged holding at high temperature leads to grain growth (Fig. 2.4e), as a result of the trend toward reduction of the system free energy, by lowering the grain boundary surface area.

The lowest temperature at which the stress-free grains appear in a specific alloy is conventionally called the recrystallization temperature. However, it must be clear that such a temperature is not uniquely defined for a specific alloy, since it depends on a wide range of microstructure-related and process-related parameters. Among the main variables of influence, we can list the following:

1. the amount of prior cold-working and its homogeneity within the workpiece volume, since the driving force for recrystallization directly depends on prior workhadening;
2. the initial grain size, which rules the amount of potential nucleation sites for recrystallization;

3. the dispersion of second phases in terms of fraction, size, shape, coherency with the matrix which also represent possible nucleation sites for the new grains;
4. the amount of recovery anticipating the recrystallization that can reduce the free energy of the system, hence the driving force for recrystallization.

The progress of recrystallization for an isothermal annealing treatment generally follows a sigmoidal trend when plotting the fraction of recrystallized material against time, as reported in the collection of data of Fig. 2.5. All the curves in the figure show an apparent incubation period before any recrystallized grain is detected, followed by an initially increasing rate of the recrystallized fraction, and an asymptotical trend when approaching the fully recrystallized condition.

Observation of the data given in Fig. 2.5 also confirms that the recrystallization kinetics obeys several laws in agreement with rules for nucleation and growth phenomena controlled by thermally activated processes and driven by the strain energy stored in the deformed metals. It is clearly demonstrated that the time for onset of recrystallization decreases as: the prior cold-working (amount of accumulated strain) increases (see Fig. 2.5a), the initial grain size decreases (see Fig. 2.5b), the annealing temperature increases (see Fig. 2.5c). Moreover, various experimental

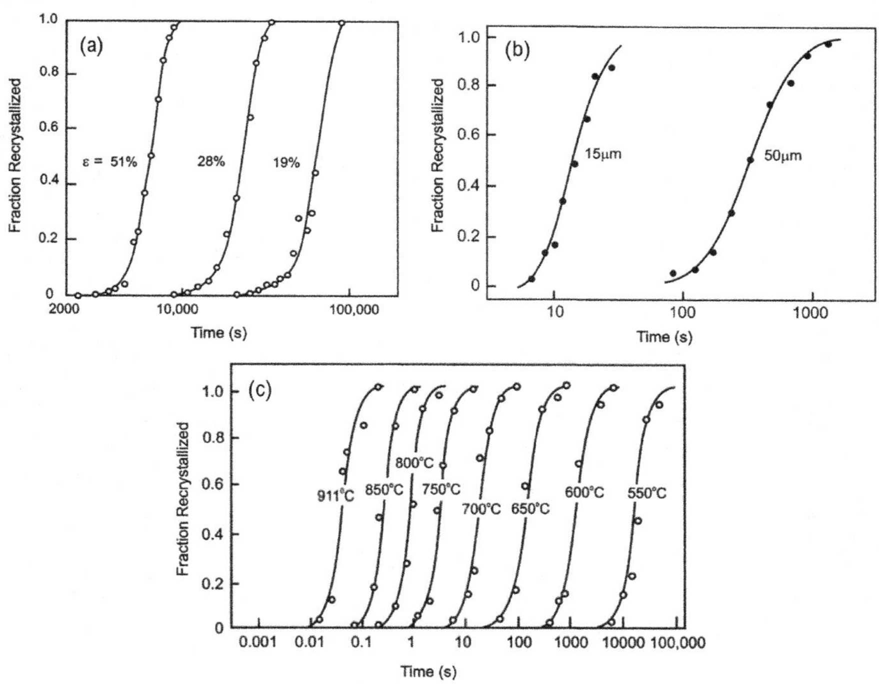

Fig. 2.5 Influence of several parameters on recrystallization kinetics. Effect of prior cold-working on aluminium annealed at 350 °C (**a**), effect of initial grain size on copper annealed at 225 °C (**b**), effect of annealing temperature on Fe-3.5%Si deformed to 60% strain (**c**). Data from (Humphreys & Hatherly, 2004)

data confirm that a minimum strain is required to start recrystallization and the size of the recrystallized grains decrease as the amount of prior cold-working increases.

2.2.2 *Practical Aspects of Cold Working*

A wide number of cold working processes exists for the manufacturing of semi-finished products. Figure 2.6 displays a representative example of a wire drawing line to help the general understanding of the cold-working cycles. Even if a detailed explanation of the technological aspects of wire drawing is out of the scope of this text, we can simply observe that a metallic wire (or a bar) is pulled from the exit side through a set of drawing dies and it is forced to plastically flow through their converging orifices, thus reducing its diameter. In general, to achieve a significant reduction of the wire diameter, the process is carried out stepwise, adopting a series of drawing stands (see inset of the figure) where the material is subjected to room temperature plastic deformation.

Wiredrawing generally starts with a feedstock made of a coil of hot-rolled rod, which is first pickled to remove any residual oxide and contaminants from surfaces and then lubricated by various methods to reduce friction when in contact with the die. As a rule, the plastic deformation achieved along each die is quite limited. It is estimated by the reduction of area (r) calculated as:

$$r = \frac{A_0 - A_1}{A_0} \tag{2.1}$$

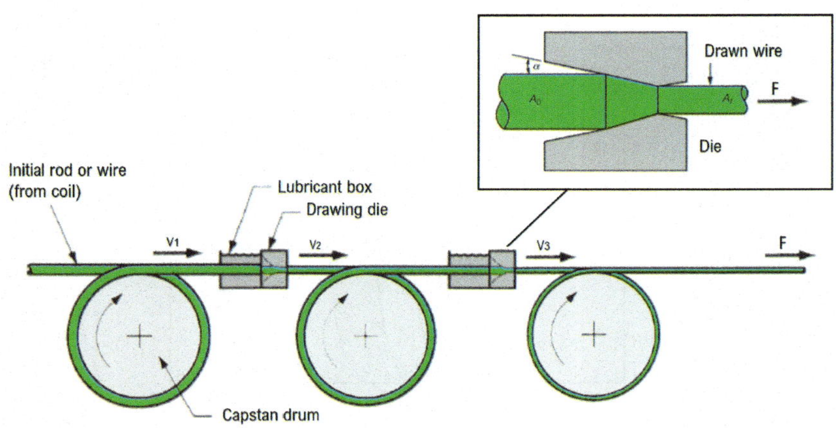

Fig. 2.6 Schematic layout of a typical multi-pass wire-drawing line

being A_0 and A_1 the initial and final cross section area of the bar, respectively. Considering the hypothesis of constancy of volume applicable to plastic deformation of metals, we can also relate the reduction of area to a corresponding true strain (ε) experienced by the metal for each drawing pass:

$$\varepsilon = \ln \frac{1}{1 - r} \tag{2.2}$$

In practical terms, the reduction of area r does not generally exceed about 30% for each pass, which corresponds to a strain of around 35.7% according to Eq. 2.2.

It has been shown in the previous section that the accumulation of plastic strain after the drawing steps produces a work-hardening effect, resulting in a stronger and more brittle material that may lead to an excessive load on the dies and unexpected failures of the workpieces. Therefore, intermediate annealing treatments are carried out on the wire to soften it and restore its original "drawability". Despite the additional costs associated to the heat treatment, an important advantage of these cold-work and annealing cycles stands in the ability of adjusting the final properties of the drawn product by tuning the amount of strain hardening, or residual strain hardening after a partial annealing, after the last deformation step.

To better define the supply condition of cold-worked semi-products, generally identified as the temper, it is in use to refer to the fraction of the maximum cold reduction sustainable for any specific material as: soft or fully annealed (temper "O"), quarter hard (temper H12), half hard (temper H14), three-quarter hard (temper H16), full hard (temper H18), extra hard (temper H19). Several other variants of this temper designations exist for specific alloys and heat treatments, which are listed in the reference standards.

Finally, it is worth discussing the actual temperature of the workpiece during the "cold-working" operation. In general, the material temperature depends on the following main factors:

- the initial temperature of the workpiece (ideally it is room temperature for the starting steps of cold-working),
- the temperature of the tools,
- the heat generated due to the plastic work,
- the heat generated due to the friction against the surfaces of the tools,
- the heat released through the tools towards the environment.

In a well lubricated process, starting from a cold workpiece, we may assume that the most important contribution comes from the conversion of the mechanical work required to produce plastic deformation into heat. Neglecting the heat losses toward the tools and the environment, the maximum temperature rise can be estimated as:

$$T = \frac{U_p}{\rho \cdot c} = \frac{\beta \cdot \sigma \cdot \epsilon}{\rho \cdot c} \tag{2.3}$$

Table 2.1 Estimated rise in temperature during adiabatic cold-working at 90% strain of several pure metals. The average flow stress during straining has been evaluated as twice the tensile yield stress to account for the work-hardening effect

Material	Yield strength (N/ mm^2)	Density (kg/dm^3)	Specific heat (J/kg °C)	Temperature rise (°C)
Al	105	2.71	897	74
Cu	210	8.96	385	104
Fe	540	7.87	449	261
Ti	220	4.51	523	159

Being β a coefficient accounting for the fraction of work converted into heat (about 0.95 for industrial cold-working processes), ρ and c the density and specific heat, respectively, while the product $\sigma \cdot \varepsilon$ represents the specific strain energy (i.e. the work of deformation per unit of material volume). Inserting typical numerical data into Eq. 2.3 and assuming to quickly cold-press by a strain of 90% cylinders made of different pure metals, their estimated rise in temperature is summarized in Table 2.1.

Although this estimate must be considered as a merely indicative evaluation, it can be noticed that the temperature rise related to the conversion of the strain energy into heat can be significant. Moreover, the actual temperature in multi-pass industrial processes may well exceed the values provided in Table 2.1 when the amount of strain energy is accumulated in a short time due to the fast strain rates.

2.2.3 Severe Plastic Deformation and Continuous Recrystallization

In previous sections it has been shown that a network of LAGBs is created when a polycrystalline metal is deformed and subsequently annealed, during the recovery stage. The formation of such substructure anticipates the stage of recrystallization which generally evolves in a discontinuous mode, namely by the nucleation and growth of the new grains starting from distinct spots in the microstructure, thus forming a set of new grains separated by HAGBs (see Fig. 2.4).

In late eighties and early nineties of the last century, it was realized that, by subjecting metals to very high strains, it was also possible to stimulate the evolution of the recrystallized structure by a gradual and more homogeneous process taking place throughout the material. The result was a fine microstructure consisting of sub-micrometric size crystallites, eventually separated by HAGBs, representing the special group of the ultrafine-grained (UFG) metals and alloys. Since no specific nucleation and growth stages could be identified in this process, the mechanism was identified as continuous recrystallization, as opposed to the above-described discontinuous mechanism which represents the rule in most industrial cold-work and annealing processes.

The key feature which determines which recrystallization mechanism is activated is that LAGBs are intrinsically unstable and spontaneously tend to nucleate new strain-free grains in a discontinuous mode when a suitable activation energy is made available. However, if sufficiently high strains are supplied to the metal under controlled conditions, then the mean LAGB misalignment can gradually increase with strain, making the microstructure more stable against discontinuous recrystallization. Eventually, with further strain reaching a total value in the range 300–700%, HAGBs can develop generating a stable microstructure according to a continuous process (Humphreys & Hatherly, 2004). Therefore, the necessary condition for the continuous evolution of the substructure is the supply of a large amount of strain at room or warm temperature (i.e. preventing discontinuous recrystallization), corresponding to the regime of the so-called severe plastic deformation (SPD). Over the years specific techniques have been developed to impart metal workpieces the required amount of SPD without incurring in sample failures due to excessive work-hardening and brittleness. Typically, large shear strains are produced by redundant methods (i.e. the deformation is imposed without any significant change in the overall shape of the specimen) and under high hydrostatic pressure. Figure 2.7 summarizes some of the most popular SPD techniques, which have been mainly developed starting from laboratory-scale approaches (Valiev et al., 2000; Azushima, 2008).

Equal channel angular pressing (ECAP) is the most used SPD processing technique (Fig. 2.7a). A billet with circular or rectangular section is pressed through a die containing two intersecting channels of the same section and it is forced to shear across the bend by an equivalent strain which depends on the angle between the channels (Φ) and the angle subtended by the outer arc of curvature of channel intersection (Ψ). Since the cross-section of the billet remains unchanged, pressing can be repeated for many cycles (N), attaining high levels of cumulative strain corresponding to the SPD regime, according to the equation (Iwahashi et al., 1996):

$$\varepsilon_N = \frac{N}{\sqrt{3}} \cdot \left[2\cot\left(\frac{\Phi}{2} + \frac{\Psi}{2}\right) + \Phi cosec\left(\frac{\Phi}{2} + \frac{\Psi}{2}\right) \right] \tag{2.4}$$

High-pressure torsion (HPT) is a further traditional process which makes use of thin disk-shaped samples subjected to high torsional strain and axial pressure under the combined torsional and compressive actions of two anvils, as shown in Fig. 2.7b. The equivalent strain at disk outer diameter can be evaluated as:

$$\varepsilon = \frac{\theta \cdot r}{\sqrt{3} \cdot h} \tag{2.5}$$

being θ the rotation angle, r the disk radius and h the disk thickness.

Accumulative roll bonding (ARB) shown in Fig. 2.7c lends itself for easier upscaling and industrialization since it uses a conventional rolling mill. By adopting a 50% reduction of section of a metal sheet, the length can be doubled at each pass. The sheet is then cut into two halves, that are degreased and brushed and then stacked together to achieve the same initial sheet thickness and length, ready to be rolled

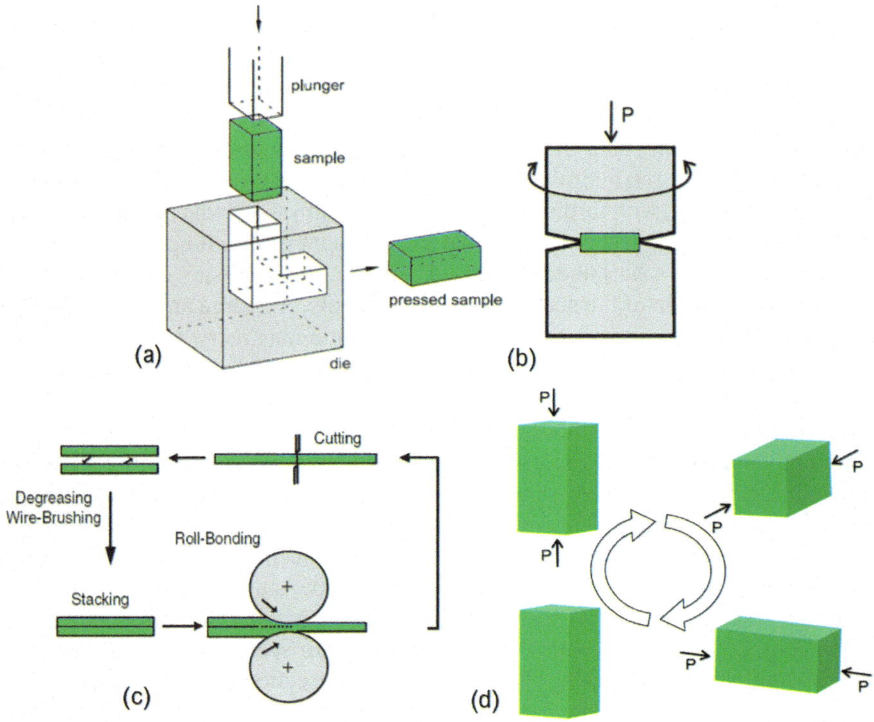

Fig. 2.7 Schematic principles of the most popular SPD methods. (**a**) Equal channel angular pressing, (**b**) high-pressure torsion, (**c**) accumulative roll bonding, (**d**) multi-directional forging

again for a further pass. This procedure can be repeated many times to reach the desired amount of deformation.

Finally, multi-directional forging (MDF) assumes that a set of sequential forgings are produced on a prismatic workpiece along its principal axes to keep the workpiece shape substantially constant as shown in Fig. 2.7d. Forging is generally carried out after a moderate preheating of the material (within the range 0.1–0.5 of the absolute melting point). The workpiece can be in principle of large size to generate semifinished products for industrial applications.

The plastic cold-working or warm-working produces a hierarchy of dislocation substructures and microstructures that have been already described in Sect. 2.2.1 for moderate strains. Figures 2.8a, b recall the initial generation of the dislocation cells and sub-grains separated by LAGBs upon cold working. Further accumulation of cold work, leading to the SPD condition, promotes the formation of thin grains stretched along the direction of the principal strain, containing the array of sub-grains with their LAGBs (Fig. 2.8c1). Eventually, the closely spaced high-angle grains become more serrated and collapse at some points, creating the ultrafine and equiaxed structure depicted in Fig. 2.8d. It is worth considering that second-phase particles can play a positive role on the fast achievement of the UFG structure, by breaking up the initial

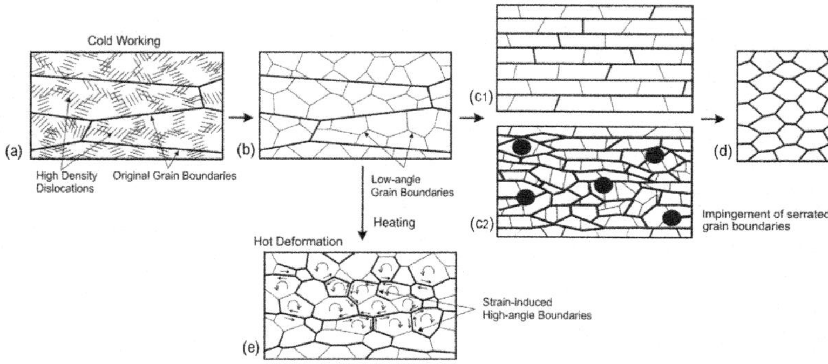

Fig. 2.8 Schematic views of the microstructures developed with increasing strain and different deformation conditions (adapted from Sakai et al., 2014)

planarity of the elongated grains and stimulating large local lattice rotations (hence formation of HAGBs) in their vicinity, as shown in Fig. 2.8c2.

Figure 2.8e indicates an alternative mechanism which is usually met during warm- or hot-working of some high-stacking fault energy materials such as the aluminium alloys. The mechanism in this case involves the gradual shift from the dislocation substructure to a stable UFG structure as a result of sub-grain rotation induced by the local lattice orientation misfit, leading to the evolution of LAGBs into HAGBs. A variant of this mechanism accounts for the dislocation trapping at the boundaries, which generates a gradual increase in sub-grain misorientation, leading to the formation of a stable UFG structure (Sakai et al., 2014).

Even considering that the specific processing technique, the deformation condition and the material tested may result in slight modifications of the illustrated mechanisms, a common aspect shared by all the SPD technologies is the ability to produce bulk metals and alloys with a submicrometric, even nanometric, grain size. Typically, the grain size of bulk UFG metals can range from around 50 to 500 nm. Moreover, the UFG structure is associated to the presence of non-equilibrium grain boundaries with higher density of dislocations and a greater number of vacancies and crystal defects than in conventionally processed metals. Second-phase particles are also subjected to a reduction in size and possible changes in local composition by virtue of the SPD processing effects.

As a result, UFG metals possess a combination of peculiar properties which attracted the attention for several applications. The first important implication of the dramatic grain refinement is the enhanced combination of strength and ductility, especially at room and moderate temperatures. The fine and equiaxed stable grain structure also makes most of the UFG metals well suitable to superplastic forming, namely the ability to sustain very high tensile elongations prior to failure, even at unusually high strain rates and low temperatures (see data in Fig. 2.9).

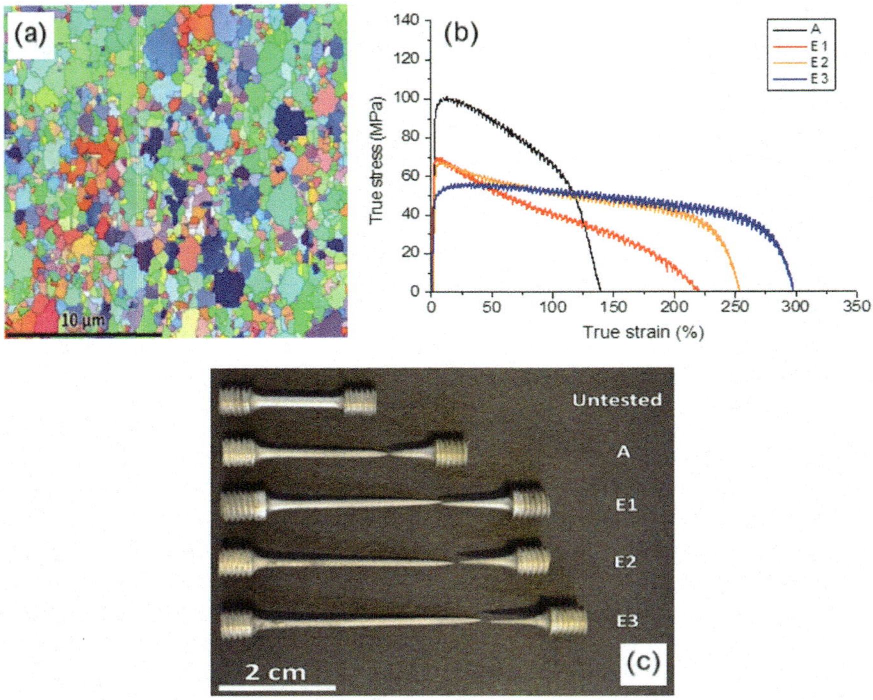

Fig. 2.9 UFG structure obtained in a ZK60 (Mg–5.3Zn–0.48Zr, wt.%) magnesium alloy by ECAP processing (**a**), and superplastic behaviour after tensile testing at 200 °C at an initial strain rate of 10^{-3} s^{-1}(**b**) and (**c**). The labels A, E1, E2 and E3 identify the starting as-extruded alloy, and the samples processed by four ECAP passes at 250, 200 and 250 °C, respectively (Mostaed et al., 2015, published with permission of Elsevier)

2.3 Hot Plastic Deformation

Hot working concerns a wide range of deformation processes carried out at high temperature. Typical hot working conditions are found in forging, hot rolling, extrusion. Some of these processes are selected as the first deformation stages for the modification of a cast billet or ingot into a wrought product.

Hot deformation is generally carried out at temperatures exceeding about 0,6 times the absolute melting temperature of the alloy, typically above 800 °C for steels, 250 °C for aluminium alloys, and 800 °C for copper alloys. However, the threshold for the term "hot" should not be considered as a unique and absolute value of temperature, but rather the result of a combination of conditions for temperature, strain rate, strain path, and material structure which enable the prompt activation of the recovery and recrystallization processes, allowing the processed material to restore its condition right during the deformation process.

As a result of this particular processing condition, large strains can be provided to the workpiece without creating any substantial strain hardening effect and consuming less deformation energy since the material is softer at high temperature. Moreover, the combination of plastic flow and elemental diffusion promoted by the high temperature allows achieving a more homogeneous elemental distribution (reducing the effects of segregation), and a refined microstructure containing a limited amount of defects due to the beneficial role of the hot compressive strains, which tend to close and weld small porosities and cracks inherited from the solidification process.

The downside of hot-working mainly consists in the reaction of the hot metal with the working environment, which is usually air. This stimulates the extensive oxidation of workpiece surfaces, with the corresponding loss of material, additional costs for finishing of the surfaces (descaling, pickling, machining), and larger dimensional tolerances of the worked semi-products. The most reactive metals and alloys can even suffer from oxygen embrittlement, while high-C steels are subjected to extensive surface decarburization due to reaction of carbon with the environmental oxygen.

2.3.1 Concepts About Dynamic Recrystallization

In Sect. 2.1, the main microstructural transformations inducing recovery and recrystallization upon annealing of a cold worked metal have been discussed. Subsequently, hot-forming has been introduced stating that this operation is carried out at a temperature such that these mechanisms can take place spontaneously and at a rapid rate, right during the deformation process (hence the notation of dynamic recrystallization) or within the short timespan during which the hot workpiece is transferred to a further deformation step. Figure 2.10 graphically summarizes the above conditions in a simplified form. The schematic drawings suggest that dynamic recrystallization conditions promote the generation of new grains right during the hot deformation process (here, when the plate crosses the roll gap), whereas static conditions are created for lower imposed strains (note the lower initial thickness of the plate in Fig. 2.10b), assuming the same deformation temperature, and can trigger the recovery and recrystallization processes only after the deformation stage, even if within a very short time in most of the practical situations.

The understanding of the alloy behaviour under hot deformation condition is of great importance for the optimization of the hot-metal forming processes. With this objective, the evolution of different alloys against recovery and recrystallization can be investigated in laboratory environment through a set of hot deformation tests (typically compression or torsion tests) performed at different temperatures, and at strain rates approaching as much as possible those encountered in the real processes. Representative hot compression curves for two alloy steels are depicted in Fig. 2.11.

The first set of curves in Fig. 2.11.a refers to a high-Cr steel deformed in the α-ferrite temperature range. Indeed, ferritic steels, along with Al alloys and α-Ti alloys are high stacking fault energy (SFE) materials typically featuring easier mobility of screw dislocations by cross slip and a higher efficiency of the recovery process during

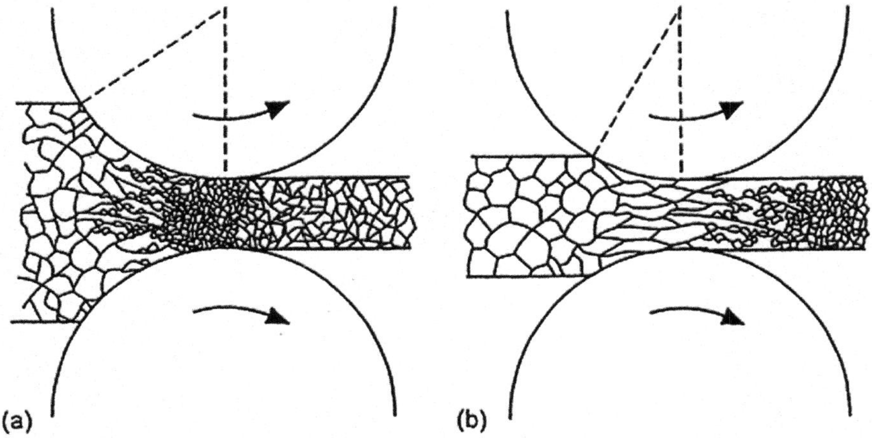

Fig. 2.10 Schematic representations of the dynamic (**a**) and static (**b**) recrystallization mechanisms created during hot working

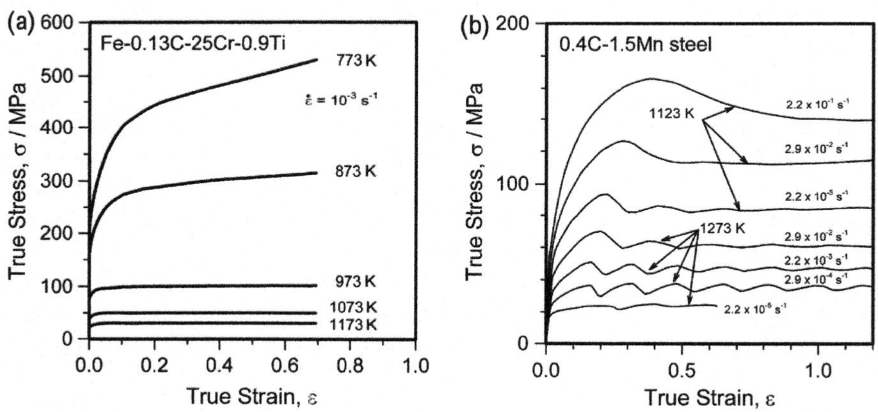

Fig. 2.11 Flow stress curves collected by hot-compression tests at different temperatures and strain rates for alloy steels in the ferrite (**a**) and austenite (**b**) temperature range (Belyakov and Kaibyshev, 1994; Sakai and Jonas, 1984, published with permission of Elsevier)

hot deformation. Accordingly, the reported flow stress curves depict an initial work-hardening stage which is counterbalanced at sufficiently high temperatures by the softening due to the rearrangement of dislocations owing to the dynamic recovery. Therefore, the high SFE materials deformed under these conditions, will mainly show the gradual elongation of their high-angle grains and a continuous evolution of the dislocation substructure which re-adapts to the increasing strain, as illustrated in Fig. 2.12.

Figure 2.11b refers to a C-Mn steel deformed in the γ-austenite temperature range. This time the example represents low or medium SFE materials (other representatives

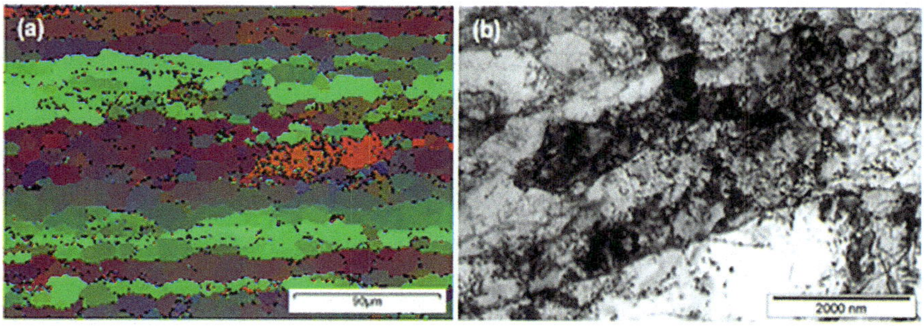

Fig. 2.12 Microstructure of a hot extruded Al–Mg–Si (EN AW6082) alloy. Orientation map image collected by EBSD (**a**) and TEM view of the sub-grain structure (**b**). In the EBSD image, the high-angle grains can be distinguished by the sharp variation of colours (green rather than blue regions), whereas the sub-grains can be identified by the slight variation of tones within the grains

are copper and Cu alloys, Mg alloys and austenitic stainless steels) that show a slightly different shapes of the flow curves. In this case, the comparatively lower mobility of the subgrain boundaries allows the build-up of a larger amount of stored energy, eventually leading to a more favourable condition for the nucleation of new recrystallized crystals at the original grain boundaries. As a result of the dynamic recrystallization process, the flow curves give evidence of a first work-hardening stage followed by a plateau and a softening phase until and equilibrium is reached between the strain hardening and the softening induced by the appearance of new stress-free grains. In some cases (see the curves at the highest strain rates in Fig. 2.11b), several peaks of decreasing height can be observed due to distinct phenomena of recrystallization of specific sets of grains in some regions of the sample.

Finally, Fig. 2.13 depicts the microstructures of two Mg alloys taken at an intermediate stages of the recrystallization process to better highlight the evolution of the recrystallization process.

2.3.2 Hot Workability of Alloys

The hot workability of metallic alloys is generally defined in terms of width of the temperature range, maximum strain rate and maximum amount of deformation which can be given to the workpiece without incurring into significant defects. On a purely schematic plot, the effects of such parameters can be visualized as shown in Fig. 2.14. As already stated, the lower limit for hot forming of a metal is the temperature which allows achieving a rate of recrystallization fast enough to avoid work-hardening and generation of defects due to poor ductility of the workpiece. As the temperature is increased, the yield strength and the work-hardening rate of materials decrease leading to improved formability. The upper limit for hot formability is theoretically achieved when reaching the melting of the alloy or, more frequently, it is ruled by

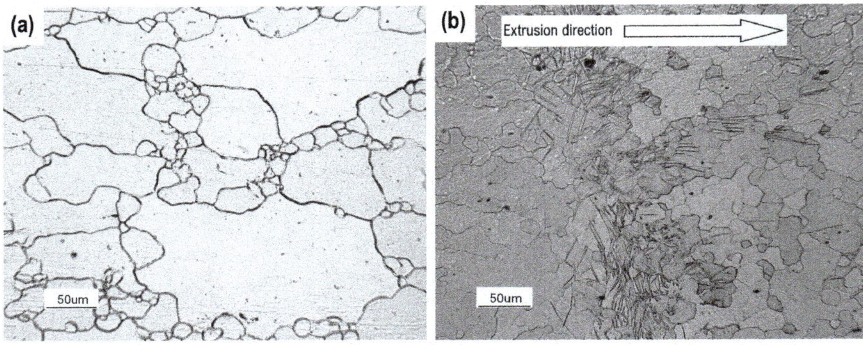

Fig. 2.13 Optical microscope image of a partially recrystallized Mg–Al (AM31B) alloy showing small recrystallized grains nucleated at boundaries of the original grains in a compression test sample (**a**); Mg-Zn (ZM21) alloy sample depicted along the section of an interrupted extrusion test where deformed and twinned grains can be recognized in the middle of the image while newly formed grains are observed on the right side, after the occurrence of complete dynamic recrystallization (**b**)

concerns about excessive oxidation, grain growth or hot shortness (also called hot cracking) of the workpiece.

It is to note that the high-temperature side of the deformation range is strongly affected by the strain rate. As the rate of deformation increases, more heat will be retained in the workpiece due to reduced time for its dissipation, approaching the adiabatic condition. This means that the pre-heating temperature should be reduced to keep it below the safety margin throughout the process.

The experimental approach adopted for the measurement of the formability curves generally relies on the evaluation of the alloy ductility (reduction of area at fracture or fracture elongation) by hot tensile tests carried out in in the temperature range of interest for the hot-working process. A schematic example of the hot-ductility curves of a structural steel is reported in Fig. 2.15. The plot considers the full range of the

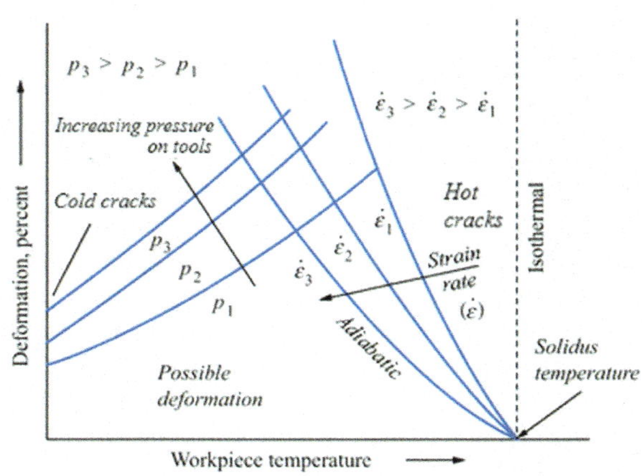

Fig. 2.14 Schematic trends of workpiece temperature, strain rate and tool pressure on formability of metal alloys

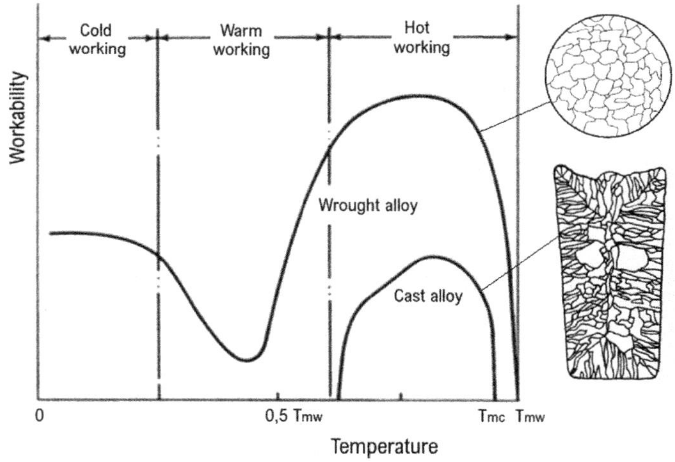

Fig. 2.15 The effect of deformation temperature on workability of cast and wrought steels

plastic deformation processing temperatures, including those for cold working, warm working and hot working.

The direct comparison between the wrought steel and its cast counterpart suggests the obvious reduced hot workability of this latter condition due to the comparatively coarser grain structure, the possible presence of casting defects and the segregation of alloying elements. It is also inferred that cold and warm plastic deformation are generally not feasible for as cast microstructures. Additionally, the upper bound of the temperature range needs to be kept conservatively below the nominal solidus temperature of the alloy due to the presence of the low-melting point constituents expected in the cast alloy. As the number of hot plastic deformation passes accumulated by the cast structure increases, the repeated recrystallization cycles and the elemental diffusion lead to a refined and more homogeneous microstructure which is more prone to hot forming. Therefore, the workability curve of the steel in the hot region would tend to rise, progressively reaching the position indicated for the wrought steel.

Considering now the left side of the plot, it can be observed that the workability curve for the wrought steel goes through a maximum within the hot working temperature range and then drops for lower temperatures. While the general trend toward reduced ductility with decreasing temperature was expected, the peculiar ductility trough found in the intermediate temperature range (from 700 to 1000 °C in low-carbon steels and high-strength low-alloyed steels) could appear as surprising. This form of embrittlement has been the subject of a large number of research projects due to important implications in many hot deformation processes and in continuous casting of steels, when the solid skin at the extrados of the curved strand is stretched in tension, inducing issues about the so-called transversal cracking (Mintz & Qaban, 2021).

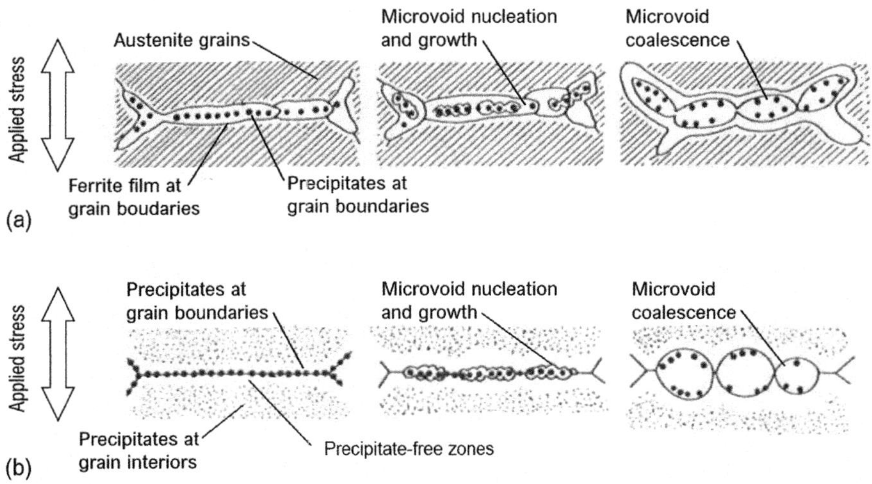

Fig. 2.16 Schematic of the intergranular failure mechanisms found in low-alloy steels deformed within the hot- ductility trough; **a** formation of hot cracks due to a soft ferrite film at austenite grain boundaries and **b** due to extensive precipitation at grain boundaries and creation of precipitate-free zones in adjacent regions (adapted from Maehara et al., 2013)

The ductility drop in the warm deformation region is particularly sharp in low alloys steels. It can be caused by the anticipated intergranular failure at the austenite grains on cooling, when a thin layer of softer ferrite appears and promotes strain localization leading to the rapid nucleation and growth of microvoids. As suggested in Fig. 2.16a, already existing precipitates (typically manganese sulphides or Nb- and V- rich carbonitrides) decorating the austenite grain boundaries can play a role on the easy generation of intergranular cracks.

Figure 2.16b depicts a slightly different mechanism, typically met in high-strength low-alloy steels. In this case, the presence of regions free from strengthening precipitates (hence softer than the interior of the grains) close to the austenite grain boundaries activates the fast nucleation and growth of voids in a way similar to the ferrite film already described.

Figure 2.17a supports the understanding of the above mechanisms. The white constituents depicted in the figure are the first fraction of ferrite which was nucleating at prior austenite grain boundaries at the testing temperature. Cracks clearly developed from those grain facets which were oriented perpendicular to the loading direction. In Fig. 2.17b one of those hot cracks is depicted by a SEM image taken at higher magnification once the strained specimen had been separated in two halves. There is evidence of stringers of precipitates located along a further grain boundary oriented perpendicular to the surface of the main crack.

Concluding the analysis of the phenomena ruling hot ductility, it can be considered that, at the left side of the trough, ductility recovers due to the presence of a larger amount of ferrite which prevents strain localization. At the right side of the trough,

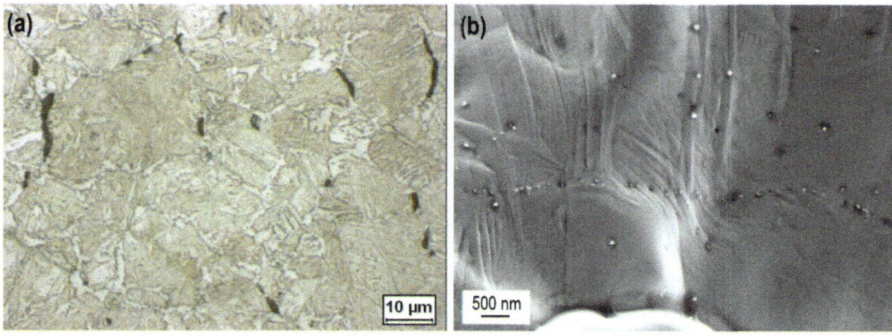

Fig. 2.17 Microstructure of a C-Mn steel strained in tension at 850 °C within the ductility trough region: (**a**) optical micrograph of a cross-sectioned specimen (loading axis was horizontal), (**b**) appearance of the face of a hot crack once the specimen has been fully broken

the high temperature enables the prompt dynamic recrystallization of the deformed structure, allowing migration of the grain boundaries from the sites where cracks nucleated and the general refining of the grain structure.

References

Azushima, A., Kopp, R., Korhonen, A., Yang, D. Y., Micari, F., Lahoti, G. D., Groche, P., Yanagimoto, J., Tsuji, N., Rosochowski, A., & Yanagida, A. (2008). Severe plastic deformation (SPD) processes for metals. *CIRP Annals—Manufacturing Technology, 57*, 716–735.

Belyakov, A., & Kaibyshev, R. (1994). Deformation mechanisms in a high-chromium ferritic steel: I. Phenomenological analysis. *Physics of Metals and Metallography, 78*, 240–245.

Dieter, G. E. (1988). *Mechanical metallurgy*. McGraw-Hill Publisher.

Hu, H. (1963). *Recovery and recrystallization of metals* (pp. 311). AIME Conference Series. Interscience Publishers.

Humphreys, F. J., Hatherly, M. (2004). *Recrystallization and related annealing phenomena* (2nd ed.). Elsevier Publisher.

Iwahashi, Y., Wang, J., Horita, Z., Nemoto, M., & Langdon, T. G. (1996). Principle of equal-channel angular pressing for the processing of ultrafine-grained materials. *Scripta Materialia, 35*(2), 143–146.

Maehara, Y., Yasumoto, K., Tomono, H., Nagamichi, T., & Ohmori, Y. (2013). Surface cracking mechanism of continuously cast low carbon low alloy steel slab. *Materials Science and Technology, 6*, 793–806.

Mintz, B., & Qaban, A. (2021). Understanding the high temperature side of the hot ductility curve for steels. *Materials Science and Technology, 37*, 237–249.

Mostaed, E., Fabrizi, A., Dellasega, D., Bonollo, F., & Vedani, M. (2015). Grain size and texture dependence on mechanical properties, asymmetric behavior and low temperature superplasticity of ZK60 Mg alloy. *Materials Characterization, 107*, 70–78.

Sakai, T., & Jonas, J. J. (1984). Dynamic recrystallizationme—Chanical and microstructural considerations. *Acta Metallurgica, 32*, 189–209.

Sakai, T., Belyakov, A., Kaibyshev, R., Miura, H., & Jonas, J. J. (2014). Dynamic and post-dynamic recrystallization under hot, cold and severe plastic deformation conditions. *Progress in Materials Science, 60*, 130–207.

Siegert, K., Möck A., & Neher R. (2000). https://aluminium-guide.com/en/talat-lectures/. Adapted from TALAT 2.0, lecture 3401—Forging alloys.
Valiev, R. Z., Islamgaliev, R. K., & Alexandrov, I. V. (2000). Bulk nanostructured materials from severe plastic deformation. *Progress in Materials Science, 45*, 103–189.

Chapter 3
Welding

Although the practice of welding by rudimentary methods dates back to the prehistoric times, it is only from the last decades of the nineteenth century that the modern welding technology was developed and accepted for the manufacturing of structures in industry. By those times, the number of welding processes increased dramatically, especially when high-intensity heat sources such as plasmas, electron beams and lasers became available.

The variety of the welding processes is nowadays so wide that even a precise definition of the term "welding" is still a matter of debate among scientific textbooks, institutional bodies and related standards. To fix ideas, one could refer to the definition of metal welding given by the BS EN 14610:2004 standard: "Operation which unifies metal(s) by means of heat or pressure, or both, in such a way that there is continuity in the nature of the metal(s) which has(have) been joined", with the additional note that: "A filler metal, the melting temperature of which is of the same order as that of the parent metal(s), may or may not be used and the result of the welding is the weld".

Therefore, it is inferred that welding involves the formation of atomic bonds at the edges of the parts to be jointed to assure the "continuity" of the weld. Continuity implies that there does not exist any distinct interface along the welded joint, both at the microscopic and macroscopic scale. This opposes to joints generated using adhesives, by bolting or by any other type of mechanical fastening, where a physical gap between the parts can always be identified.

The welding processes can be broadly classified into those that achieve the continuity by melting the edges to be joined (hence the name of fusion welding) and those that produce the bonds by plastic deformation and/or by solid-state diffusion through the action of pressure, with or without the combined effect of temperature (hence the term of pressure welding or solid-state welding). Finally, there exists a further group of the brazing and soldering processes where fillers with lower melting points than the base metals are used to generate the material continuity. A proper heat source

M. Vedani, *Metal Science in Modern Manufacturing Technologies*, Springer Series in Materials Science 351, https://doi.org/10.1007/978-3-031-97125-9_3

is selected to promote melting, infiltration and re-solidification of the filler material between the edges to be joined, still keeping the base material in the solid state.

This chapter will mainly be focussed on the metallurgy of fusion welding. However, in the first section, a general overview about the main welding processes will be given to better hook practical matters to the scientific concepts to be discussed in the following sections.

3.1 Welding Processes

3.1.1 Fusion Welding Processes

Within the fusion welding processes, a classification is usually made based on the source of the energy required to melt the material (Messler, 1999, 2011). We can list processes driven by chemical sources, where the energy is provided either by the combustion of a fuel gas which combines with oxygen, or processes relying on exothermic combustion synthesis reactions of powdered metals with carbon or other oxides. Figure 3.1 schematically shows the principles of the oxyfuel gas welding technique, selected as one of the most popular processes using a chemical energy source. In this case, the energy is produced by the reaction of a hydrocarbon gas (typically acetylene, C_2H_2) with oxygen, according to the two reactions of primary and secondary combustion, respectively:

$$C_2H_2 + O_2 > 2CO + H_2 + 448\,\mathrm{kJ/mol} \tag{3.1}$$

$$2CO + H_2 + 1/2O_2 > 2CO_2 + 2H_2O + 812\,\mathrm{kJ/mol} \tag{3.2}$$

By adjusting the relative flows of oxygen and acetylene, it is possible to change the chemical nature of the welding flame, keeping it neutral or causing a preferred reduction rather than oxidation on the surface of the workpiece, Oxidative conditions are preferred to clean (burn) grease and other contaminants from surfaces and to limit the loss of the most volatile elements from alloys. Reducing conditions can be beneficial to remove the oxide film from surfaces before welding, especially for the most reactive materials such as aluminium and magnesium alloys,

A second extremely flexible source of energy for fusion welding is represented by the electric arc, which is produced between an electrode and the conductive workpieces to be welded once they are inserted in an electric circuit and separated by a gap. An electric arc is an electrical breakdown formed at the circuit gap, that produces a prolonged electrical discharge (hence significant energy transfer) provided the applied electrical potential is high enough. The arc consists of negatively charged electrons that move across a normally nonconductive medium such as the air or any other gas found in the circuit gap. Electrons are extracted from an electrode (the cathode) and attracted by the positive anode. Additionally, these fast-flowing

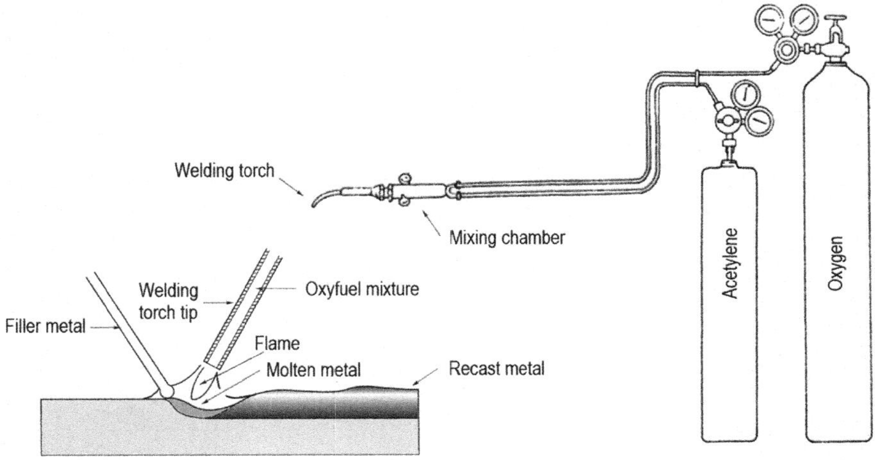

Fig. 3.1 Schematic of an oxyfuel gas welding process, highlighting the torch tip and the welding pool

electrons can impact against the gas atoms, producing the ejection of further electrons and the generation of positively charged ions, which will move toward the cathode under the effect of the same potential.

The electric arc can be created using either direct or alternate current. In case of direct current, the constant polarity can be set to have the workpiece to be welded positive (anode) with respect to the welding electrode (cathode), thus operating under the so-called direct current straight polarity. This opposes to the direct current reverse polarity mode where the workpiece is negative, the electrons are attracted toward the welding electrode and the ion current strikes the workpiece. When an alternate current is used, the polarity is repeatedly reversed at each cycle and the net effect on the welding operation can be roughly considered as a combination of the straight and reverse polarities. The power provided by the electrical source is used to heat and melt the joint edges through the action of the kinetic energy of the electrons and ions. Electrons possess higher kinetic energy since they can be accelerated at much higher speeds than ions, as a result the direct current straight polarity is more effective in transferring heat to the workpiece to be welded. On the contrary, the reverse polarity mode, featuring heavier ions striking the workpiece, is preferred when a strong cleaning action is required for the parts to be welded.

In Fig. 3.2 several schematics about the most used electric-are welding processes are collected. The main differences among the processes stand in the methods by which two main technological requirements are fulfilled. Firstly, it must be recalled that most of the engineering metals and alloys in the molten state are extremely reactive with the environment. Therefore, they must be protected from air oxidation by a suitable inert gas shield or by a slag covering the molten weld pool. Secondly, a filler metal is often required to increase the volume of the molten weld pool and to modify its overall composition by dilution with the base metal. In this perspective, there exist two broad arc-welding groups: those relying on consumable electrodes, which provide the arc and the filler metal as they are consumed during melting, and those based on non-consumable electrodes (usually made with high melting-point metals such as tungsten), where the filler, if any, is separately added as a wire or a rod to the molten weld pool.

Figures 3.2a, b depict two representative welding processes using consumable electrodes called shielded metal arc welding (SMAW) and gas metal arc welding (GMAW), respectively. In SMAW, metal electrodes in the form of sticks of finite length are clamped into a manually handled electrode holder, which is connected to

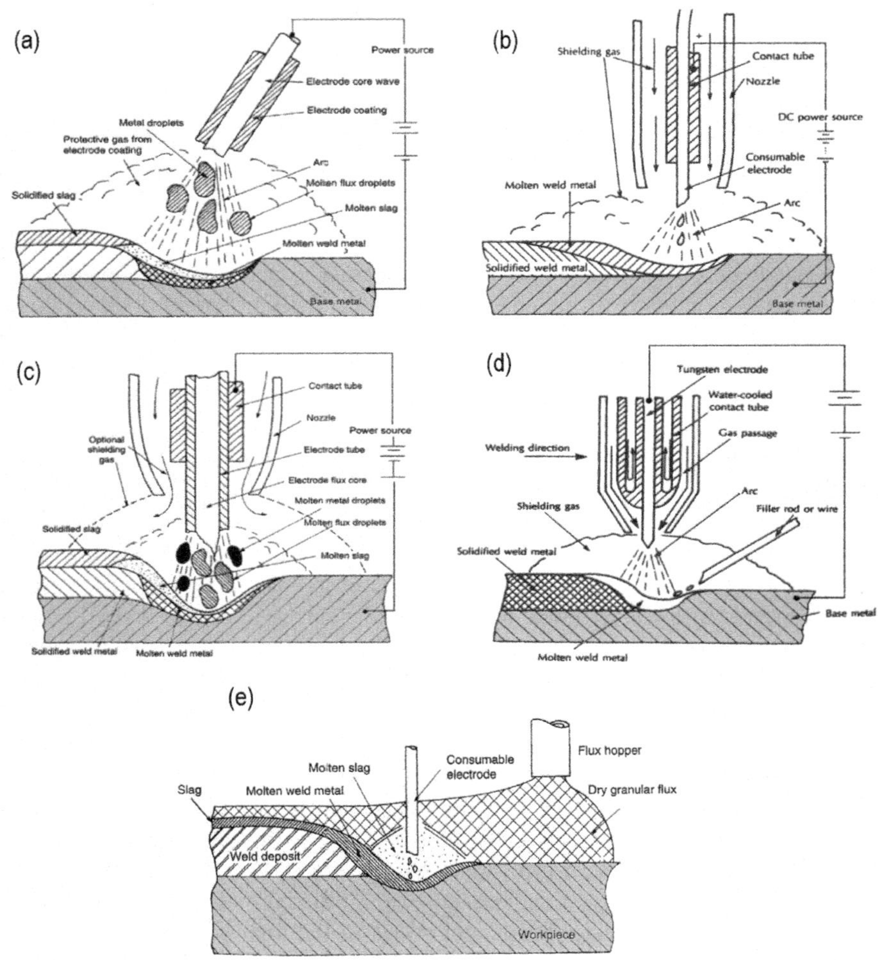

Fig. 3.2 Schematics of some popular electric-are welding processes; (**a**) shielded metal arc welding, (**b**) gas metal arc welding, (**c**) flux-cored arc welding, (**d**) gas tungsten arc welding, (**e**) submerged arc welding (Messler, 1999, published with permission of Wiley-Interscience)

the power source. The electrodes are covered with a flux which decomposes during melting, forming a protective gas cover to the molten pool and producing a molten slag that metallurgically refines the molten metal and covers the melt pool to further protect it from environmental contamination. On the contrary, in GMAW the filler consists in a continuous wire which is fed through a torch where it becomes the consumable electrode. Protection of the melt pool from the environment is assured by a shielding gas which can be totally inert or semi-active, with small addition of oxidizing rather than reducing agents (e.g. H_2 or O_2) to fulfil specific needs during the welding operation.

In Fig. 3.2c the flux-cored arc welding (FCAW) process is depicted. To increase productivity with respect to the SMAW, the consumable electrode now takes the form of a continuous tubular metal wire (the filler metal) which contains a core of the granular flux with the same functions of the electrode coatings in the SMAW process.

The submerged arc welding (SAW) shown in Fig. 3.2e is a last example of a process using consumable electrodes. The filler metal is supplied as a continuous wire fed into a torch, which is usually driven automatically. Very accurate protection from environment and increased power efficiency are made possible by covering (submerging) the whole area of the weld with a granular flux. Once the welding operation is started, the arc and the molten pool are totally shielded by the molten flux and the remaining unmolten powder. As a result of this protection, the energy losses due to convection and irradiation are dramatically reduced and the deposition rates of the filler metal can become very high, SAW processes are also selected for the production of thick cladding layers by virtue of these peculiar advantages.

Among the welding processes using non-consumable electrodes, gas tungsten arc welding (GTAW) represents the most popular technology. The basic principle of GTAW is shown in Fig. 3.2d. A permanent tungsten electrode is used to establish an arc with the workpiece to be welded and a water-cooling system is adopted to avoid heat accumulation in the torch, which is usually operated manually. The process can be performed with or without the support of a filler metal which is provided as a stick fed to the weld pool from a lateral position. A shielding gas flow delivered coaxially to the torch enables the protection of the hot electrode and of the molten pool. Argon, helium or mixtures of the two gases are generally used. Argon, being heavier than air, offers improved shielding efficiency since it stays in the weld area. Helium in turn produces a hotter arc, hence increased transfer of energy to the weld.

Finally, it is worth mentioning that there exists a variant of the GTAW, making use of a plasma instead of the electric arc. In plasma arc welding (PAW) an argon flow is conveyed into the torch and becomes ionized by the potential created between a tungsten electrode and the inner part of the nozzle (the so-called primary arc), generating the plasma. The plasma can then be redirected (as a transferred arc) to the welding area by setting a suitable potential between the torch and the workpiece itself. The advantages of PAW compared to GTAW can be summarized in the increased intensity and the energy density of the plasma source which translates into deeper penetration capability and faster welding speeds.

The ability to produce a high energy–density beam is also distinctive of other processes such as electron beam- and laser beam welding (EBW and LBW, respectively). In EBW, a beam of electrons produced by thermionic effect from a cathode in a gun is strongly accelerated by high potentials (of the order of 10–100 kV) and directed to the workpiece placed in a vacuum chamber to limit the scatter of electrons with environmental gas molecules. In LBW, a dense beam of photons is used, which will be travelling at the speed of light to the target.

In these processes, the extremely high energy of the concentrated beams is instantaneously converted into heat once it impinges on the metal surface. The resulting heating rate is generally so fast that that the temperature at the spot of beam interaction with the surface can quickly rise, exceeding the melting point and even the boiling point of the material. Figure 3.3 depicts two possible interaction modes of a beam with a material depending on its energy density. In electric arc welding processes and in PAW operated at moderate energy density, the workpiece is heated by conduction from the surface inwards, hence the name conduction mode (Fig. 3.3a). When the beam energy density is increased, reaching values in the range of 10^9–10^{10} W/m^2, typically in PAW operated at high energy, EBW and LBW, a shift to the so-called keyhole mode becomes evident (Fig. 3.3b). The term keyhole refers to the action of the vapour cavity that forms due to the intense heating during welding. It has been demonstrated that the moving beam creates a deep moving cavity of vapour which is surrounded by an annular region of liquid metal, dynamically kept in position by the vapour pressure and the surface tension. The beam travel allows the liquid region to rejoin at the trailing edge of the weld pool, generating deep and narrow weld geometries, even for thick metal plates.

An additional source for the welding heat can also be provided by the Joule effect, exploited by the wide group of the resistance welding processes. An electrical current flowing in a circuit produces heat losses that depend on $I^2 \cdot R$, being I the current intensity and R the local resistance in the circuit. Therefore, the Joule heating can be used to locally melt and weld parts that are inserted into an electrical circuit, as depicted in Fig. 3.4. In particular, the picture schematically shows the principle of the resistance spot welding: two overlapped metal sheets are clamped with pressure into a circuit using a pair of water-cooled copper electrodes. The electrical resistance generally reaches two relative peaks at the contact interfaces between the electrodes and the metal sheets, but the absolute peak of resistance is attained at the interface between the two metal sheets. When a proper amount of current is circulating, heat generation and melting will start from that interface, producing a weld nugget, whose typical microstructure is reported in Fig. 3.4b.

Resistance spot welding is widely used in the automotive and electric household appliance sectors for joining structural metal sheets and other parts. As an alternative to the spot welding, there exist several other types of resistance welding configurations. For instance, seam resistance welding processes can generate continuous welds, projection resistance welding is useful to better shape the geometry of the nugget starting from predefined profiles created in one of the sheets to be welded. For further details about these technologies, the reader is referred to specialized textbooks (Kou, 2003; Messler, 1999, 2011).

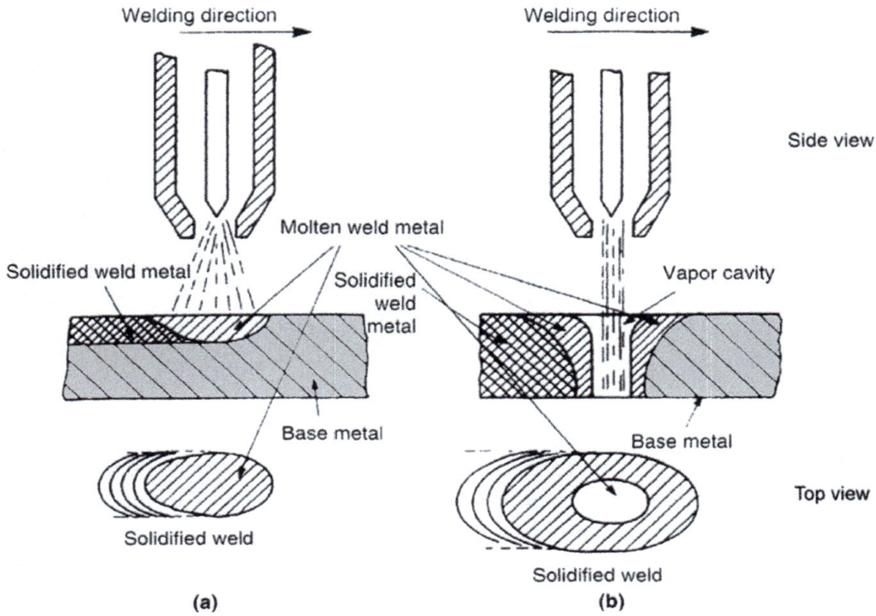

Fig. 3.3 Beam-metal interaction in high energy–density beam processes; (**a**) conduction mode, (**b**) key-hole mode (adapted from Messler, 1999)

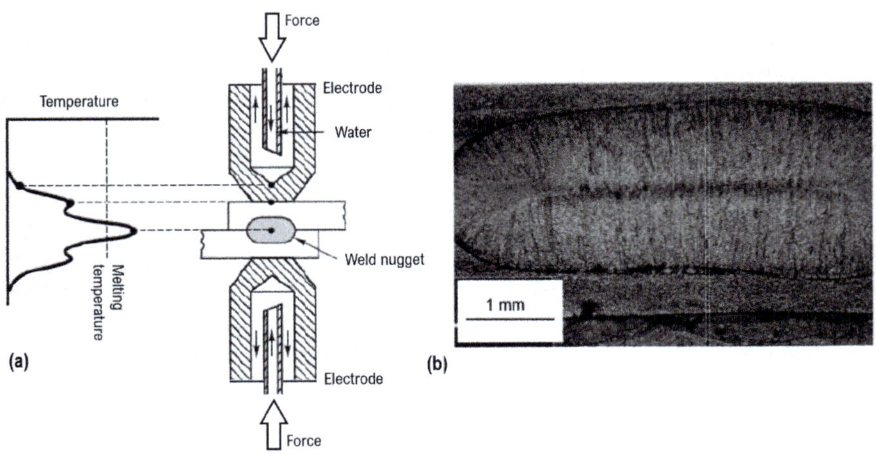

Fig. 3.4 Schematic of the resistance spot welding process (**a**) and representative micrograph of a weld nugget (**b**)

Fig. 3.5 View of a circuit board with connection realized by soldering

Brazing and soldering are further processes which also involve the use of a molten metal. However, the main difference when compared to the previous fusion welding processes stands in the use of a relatively low-melting filler metal for the formation of the welded joint. Even though in the industrial practice the distinction between brazing and soldering is frequently unprecise or blurred, brazing formally refers to processes using filler metals with a melting point exceeding about 450 °C while soldering refers to filler metals with lower melting temperatures. One of the most familiar soldering processes consists in the use of tin (melting point of 231.9 °C) which is easily melted by a hot metal tip and deposited/infiltrated as small droplets to create stable electrical connections of the electronic components on circuit boards, as shown in Fig. 3.5.

In brazing and soldering the formation of the bond takes place by the spontaneous infiltration, driven by capillarity, of the molten filler metal in the gaps created within the edges of the solid base metals, as shown in the schematic image of Fig. 3.6a. Representative micrographs of the microstructure generated in a stainless steel sheet assembly brazed with copper are also displayed in Fig. 3.6b, c. It is confirmed that the base metal sheets do not macroscopically melt and keep their original thickness and shape, while the brazing alloy infiltrates the small gap left between the two sheets creating the joint. However, a closer view of the microstructure (Fig. 3.6c) also demonstrates that an effective metallurgical bond is created at the filler-metal/base-metal interfaces by mutual reaction and small-scale diffusion between the two materials involved.

Brazing and soldering are carried out by many forms of technological processes which all share some common steps consisting of (i) cleaning the surfaces and removing oxides from the parts to be joined, (ii) arranging the parts so that the joint gap for filler metal infiltration is correct at the brazing temperature, (iii) heating the full assembly or the joint area, the filler metal being provided while heating or pre-placed into the joint gaps (as foil or paste) prior to the heating stage, (iv) cooling of the assembly after the filler metal has properly infiltrated. Heating in turn can be performed using concentrated heat sources such as flames (often manually operated)

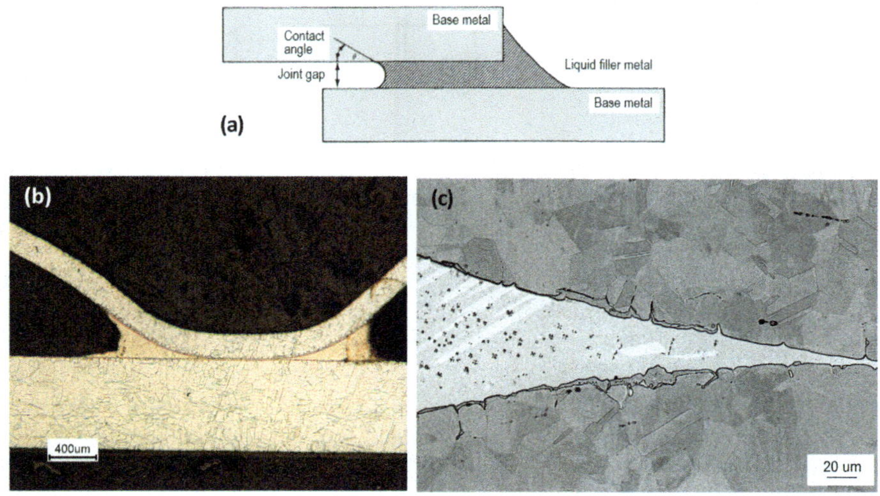

Fig. 3.6 Schematic of a joint obtained by brazing (**a**) and representative micrographs of copper-brazed type 316 stainless steel sheet joint (**b** and **c**)

or specifically shaped resistance and induction heaters. For the production of large batches, vacuum brazing furnaces are used where the full assembly is introduced, together with the pre-placed filler metal and held at the suitable brazing temperature (Roberts, 2013; Way et al., 2020).

3.1.2 Non-fusion Welding Processes

Opposed to the previously described processes, there exist the non-fusion welding processes by which bonding of the materials to be joined is achieved by plastic deformation and solid-state diffusion. We can classify three main groups of non-fusion processes: (i) pressure welding processes, that make use of pressure to induce extensive macroscopic deformation to the workpieces; (ii) friction welding processes, that exploit frictional effects produced by the relative movement of the edges to be joined in combination with pressure to induce microscopic deformation at the interface; (iii) diffusion welding, where only (or at least predominantly) solid-state diffusion activated by the high temperature is responsible for the generation of the bonds between the parts.

Pressure welding can be performed at room temperature or at high temperature. In the former case it is mainly applied to ductile metals (typically soft face-centred cubic metals like Al, Cu, Au, Ag) that can be extensively deformed to produce rigid joints without any significant intermixing. Hot pressure welding can be applied on a broader range of metals and alloys but requires vacuum chambers or inert gas protection to prevent surface oxidation which would otherwise impair the metallic continuity

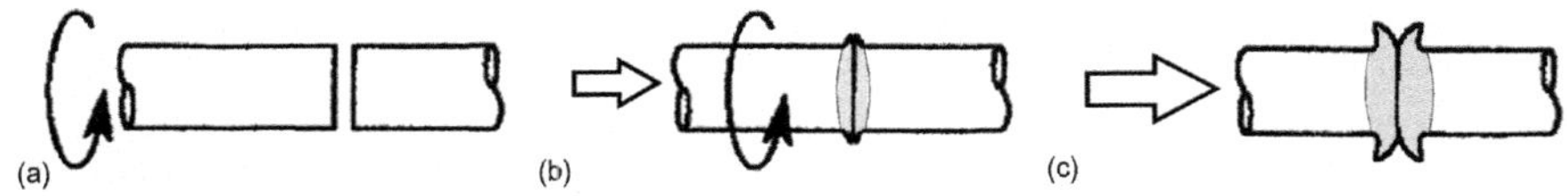

Fig. 3.7 Schematic steps of the friction welding process consisting of setting a relative angular/linear movement between parts (**a**), applying a pressure to produce heat by friction (**b**), upsetting at higher pressure the hot-metal edges once the relative movement has stopped (**c**)

across the welded joint. For this latter variant, several processes are available such as pressure gas welding, forge welding, roll welding, explosion welding.

In friction welding, the mechanical energy produced by the friction of two parts in relative motion under an applied pressure is converted into heat, which allows establishing the desired metallurgical continuity. Figure 3.7 describes the main steps involved in this process. It must be highlighted that the continuity of the joint is mainly achieved due to the final application of an additional pressure under a static condition on the hot edges, once the relative motion has stopped.

A recent process development still belonging to the group of friction welding is represented by the friction stir welding. This technology has been originally patented at TWI (The Welding Institute, UK) in 1991. It makes use of a fast-spinning tool which is pressed against- and translates along the interface of two abutting plates. Figure 3.8 illustrates the principle by which the welded joint is produced. Frictional heating mainly generated by the sliding of the tool shoulder against the surfaces of the joint induces the softening of the edges and the stirring action of the material by extensive plastic deformation, resulting in the generation of a continuous joint. The friction stir welding has the advantage of producing high quality welds with excellent mechanical properties and limited distortions. Moreover, for simple geometries of the parts, it can be easily automated for high volume productions at fast welding speed.

Finally, diffusion welding (also known as diffusion bonding) mainly relies on solid-state atomic diffusion for the generation of the bond across a clean interface of two metallic parts. Moderate pressure is still applied to these processes, but it is only used to enable the intimate contact of the surfaces, without any need to provide plastic deformation at the macroscopic scale. Diffusion welding offers the advantage of forming solid state joints even with dissimilar alloys and different material groups (e.g. metal to ceramics or even ceramics to ceramics), with limited effects otherwise given by localized heating.

The typical evolution of the microstructure across the interface during diffusion welding is depicted in Fig. 3.9a–d. It is shown that after a first stage of asperity contact, the slight applied pressure produces a distinct interface boundary, which subsequently tends to dissolve by grain boundary diffusion. Concurrently, the interface porosity is also reduced till full continuity and integrity are achieved across the joint.

In Fig. 3.9e an example of the microstructure produced by diffusion bonding of a 316 stainless steel to a free-machining 303 stainless steel (rich of non-metallic

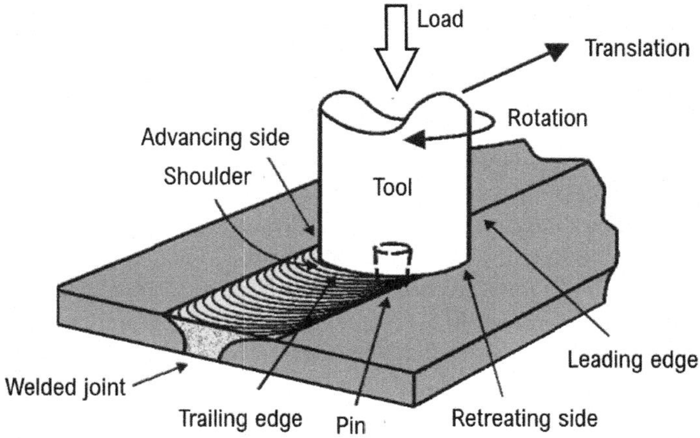

Fig. 3.8 Schematic of the friction stir welding process

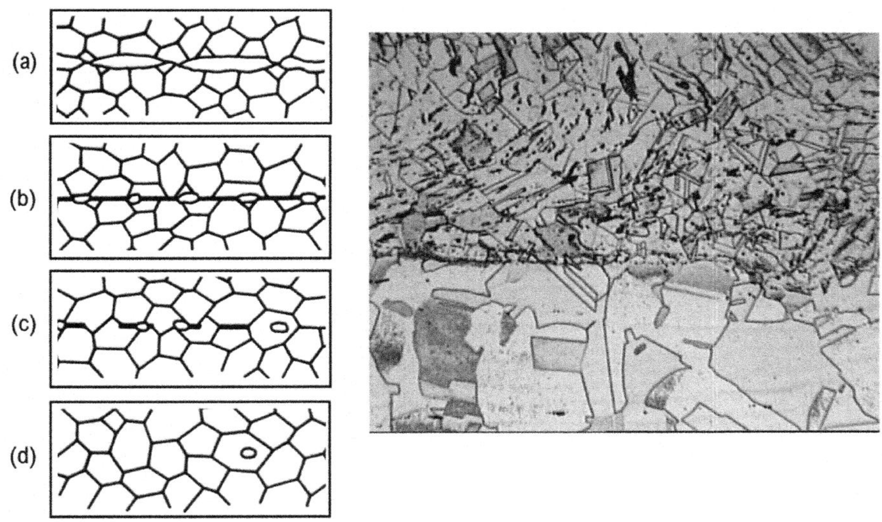

Fig. 3.9 Evolution of the interface microstructure during a diffusion welding process (**a**–**d**); microstructure of a diffusion weld created between a 316 stainless steel in the lower part of the image and a 303 stainless steel (**e**)

inclusions) is shown. The image recalls the continuity of the microstructure across the joint, with grains that extend from one side to the other of the interface.

3.2 Structure of Fusion Welded Joints

Considering the practical importance of fusion welding and the direct link with the fundamental solidification concepts provided in Chap. 1 of this book, particular attention will be paid now on the solidification mechanisms of fusion welded joints and on the development of their peculiar microstructures.

Figure 3.10 shows the general aspect of a butt weld formed by placing two plates of metals end-to-end and then welding them along the joint line. The weld bead is the physical evidence of the joint, formed by the so-called weld metal which is subjected to the melting and solidification cycles induced by the welding process. The weld metal zone is created by the metal volumes corresponding to the edges of the plates involved in melting and the additional filler metal, if any, used to increase the overall volume of the weld bead.

Adjacent to the weld metal zone, separated by the fusion line, there is the heat affected zone (HAZ), corresponding to those portions of the joint that remain at the solid state throughout the welding process, but are subjected to a significant rise in temperature due to the welding heat input. The HAZ is located on both sides of the weld metal zone and features microstructural and strength modifications owing to these, often unwanted, welding thermal cycles. Eventually, at a greater distance from the weld zone, we reach the unaffected base metal (also called parent metal) showing the properties of the original metal plates.

3.2.1 Solidification of Fusion Welded Joints—The Weld Metal Zone

The weld metal zone can be considered as the result of a moving molten pool created by the heat source, be it an oxyfuel gas flame, an electric arc or a high energy–density beam, which progressively melts new portions of the base metal at its leading edge

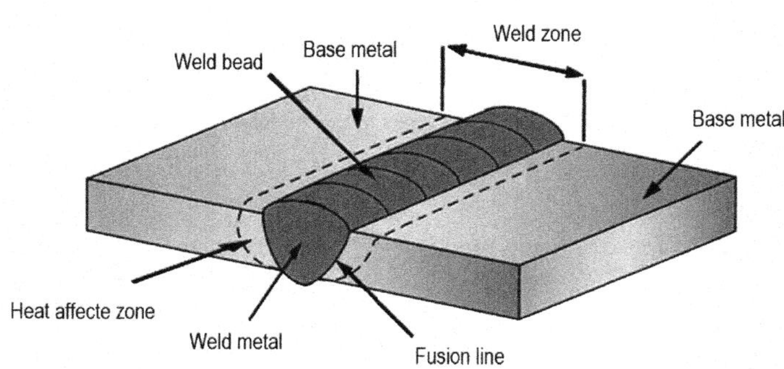

Fig. 3.10 General structure of a fusion welded butt joint

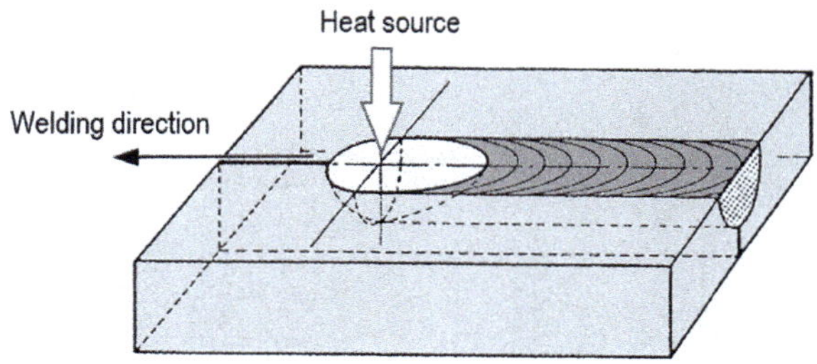

Fig. 3.11 The generation of a weld bead by the movement of the heat source along the welding direction

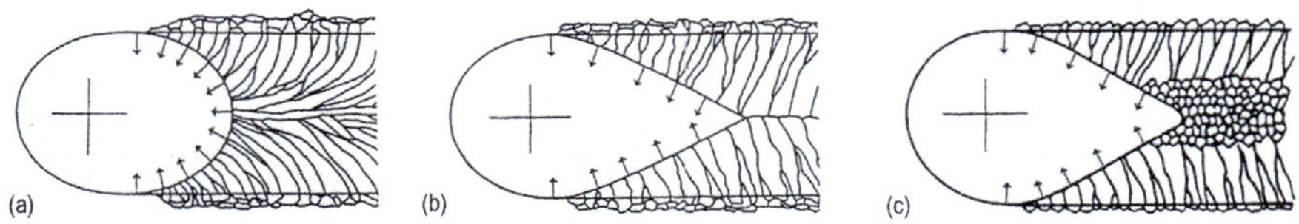

Fig. 3.12 Top views of the melt pool showing the growth direction of grains in the weld metal zone. Elliptical melt pool shape (**a**), tear-drop melt pool shape (**b**), tear-drop melt pool shape with equiaxed grains along weld centerline (**c**)

and allows the re-solidification of the weld metal at its trailing edge, as schematically shown in Fig. 3.11.

Depending on the travel speed of the heat source, the melt pool can assume an elliptical shape for low-to-moderate speeds (Fig. 3.12a) or a tear-drop shape for higher travel speeds (Figs. 3.12b, c).

It can be readily understood that the weld-metal zone is a matter of a small-size casting where the (moving) mould is represented by the solid-substrate grains facing the melt pool. By virtue of the matching crystal structure, the existing base metal grains exposed to the fusion line will act as nucleation surfaces for the weld metal grains, which will easily grow keeping the same crystallographic orientation and extending the grain boundaries of the substrate, as shown in Fig. 3.13. This mechanism is known as epitaxial nucleation.

Once the grain will extend into the weld metal zone, away from the fusion line, their growth direction will be progressively deviated according to the maximum temperature gradient and the main direction of heat extraction, namely the direction perpendicular to the melt pool boundary (shown by the small arrows in Figs. 3.12 and 3.13a). However, it is also known that there exist crystal directions leading to the fastest growth rate (for the face-centered and body-centered cubic unit cells this is the < 100 > direction). Therefore, during their growth, the grains possessing their preferred growth direction perpendicular to the melt pool boundary will grow more easily than those less favourably oriented, leading to a competitive growth mechanism within the weld metal volume (Fig. 3.13c).

The above considerations can also be used to justify the difference between the substantially straight columnar grains formed in tear-drop shaped melt pools of

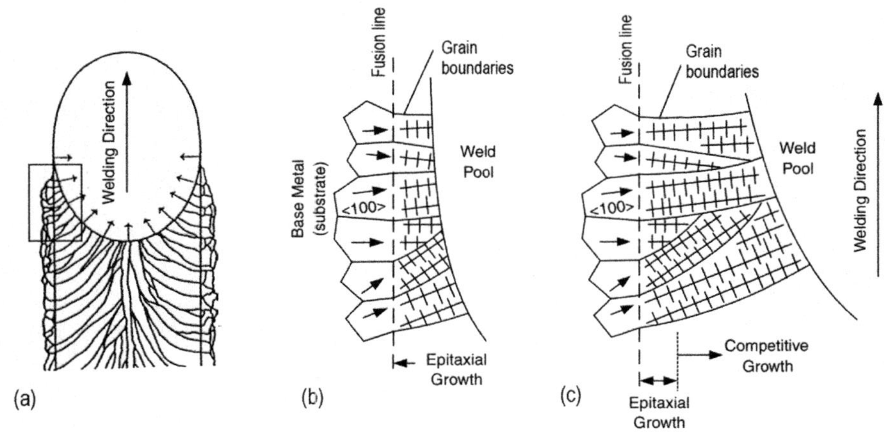

Fig. 3.13 Schematic (top views) of the development of grains in the weld metal (**a**) and enlarged view of the area in the box showing the epitaxial nucleation and competitive growth (**b** and **c**)

Fig. 3.12b, and the curved trend found in welds with elliptical-shaped melt pools shown in Fig. 3.12a.

Figures 3.14 and 3.15 depict two examples of the typical microstructure generated in fusion welds close to the fusion line. The epitaxial nucleation mechanism of the weld metal grains (on the right side in the figures) is readily recognized but it can additionally be observed that different solidification microstructures are produced in the weld metals: planar and cellular rather than columnar grains can be recognized in these two examples.

It has already been discussed in Chap. 1 (Sect. 1.1.3) that, depending on the temperature gradient in the liquid and on the growth rate of the solid front, different extensions of the constitutionally undercooled zone ahead of the solid/liquid interface

Fig. 3.14 Optical micrograph taken across the fusion line (highlighted by the white line) of a 6082 aluminium-alloy GTAW joint

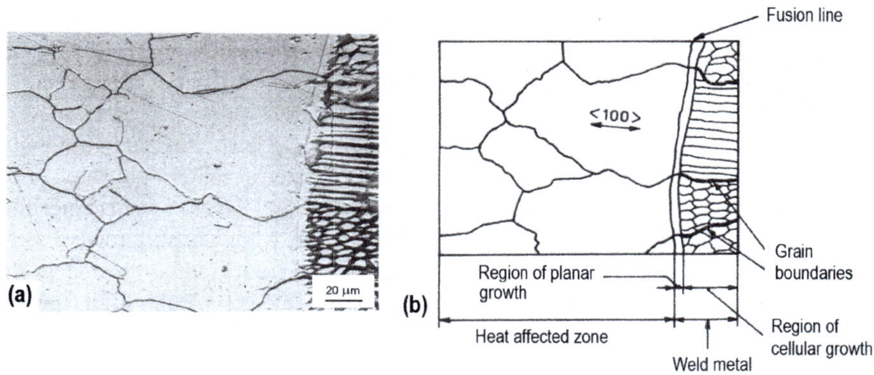

Fig. 3.15 Optical micrograph taken across the fusion line of a 316 stainless steel SAW joint (**a**) and sketch of the specific regions observed (**b**) (adapted from Piatti & Vedani, 1989)

can be generated. Correspondingly, specific microstructures are formed (refer to Fig. 1.16).

A well-known chart qualitatively summarizing the effect of temperature gradient (G) and growth rate (R) on the solidification mode of alloys is reported in Fig. 3.16 (Kou, 2003). It is inferred that the ratio G/R mainly determines the morphology of the solidification microstructure: planar, cellular, columnar-dendritic, equiaxed-dendritic, while the product G·R defines the size of the microstructure: finer rather than coarser structures.

As a matter of fact, one of the most peculiar aspects about solidification in welds is that, even within the same weld pool, different local solidification conditions can be found. A qualitative comparison between the solidification occurring at the fusion

Fig. 3.16 Effect of temperature gradient and growth rate on solidification mode (adapted from Kou, 2003)

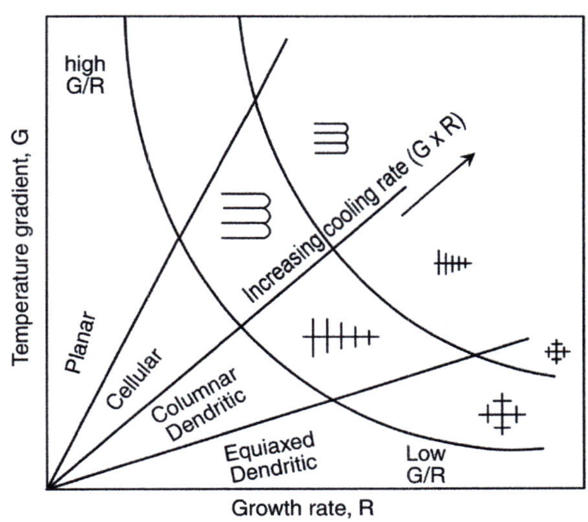

line (at the very first stages of nucleation, where new grains grow from the HAZ substrate) and next to weld centerline (where the two fronts coming from opposite sides impinge on each other) suggests that the growth rate R reaches the maximum value, matching the welding speed, in the latter position whereas it drops to very low values, theoretically to zero at the fusion line. As far as the thermal gradient is concerned, considering that the peak temperature is located at center of the melt pool, it can be assumed that G will be higher at the fusion line and lower at weld centerline. Therefore, we can suppose that the solidification front, while shifting toward weld center and deviating along the weld axis is subjected to a gradual change in the solidification conditions. At the fusion line, the G/R value is typically high (top left position in the chart of Fig. 3.16) and planar or cellular microstructures are expected. When approaching the weld centerline, the G/R ratio tends to shift to the bottom right position in the diagram, indicating the preferential formation of columnar- or equiaxed-dendritic grains.

Finally, it is worth quantifying the amount of energy used for the welding operation and the sources for its dissipation to the surrounding environment. The net heat flow delivered to the weld is usually evaluated in terms of heat input per unit of linear length of the heat-source path. It can be practically calculated as:

$$H_{net} = \eta \frac{P}{v} \tag{3.3}$$

Being H_{net} the net energy input (in J/mm), P the heat source power (in W) and v the welding speed (in mm/s). An efficiency parameter η ranging from 0 to 1 is also included in the formula to account for the thermal energy losses upon energy transfer from the heat source to the workpiece and within the workpiece material. Figure 3.17 graphically summarizes the main losses of energy in a fusion-weld process. It is demonstrated that the heat transfer efficiency is affected by radiation losses, by convection of heat to the surrounding environment and to the shielding gas, and by conduction to the electrode and consumable filler wire, or to the mass of the parent metal surrounding the weld. This latter heat flow is responsible for the welding thermal cycles induced on the solid metal during the welding process, leading to the microstructural modification found in the HAZ.

We can conclude this preliminary discussion recalling the idea that, for any process that employs a heat input to generate a permanent weld, there will be an effect on the microstructure and properties, even outside of the weld metal, which becomes mostly evident in the HAZ. Moreover, in multipass welds, where the weld bead is created through the deposition of more than one welding track, the different welding thermal cycles will accumulate and overlap each other with a limited time delay, generating even more complex and extended effects.

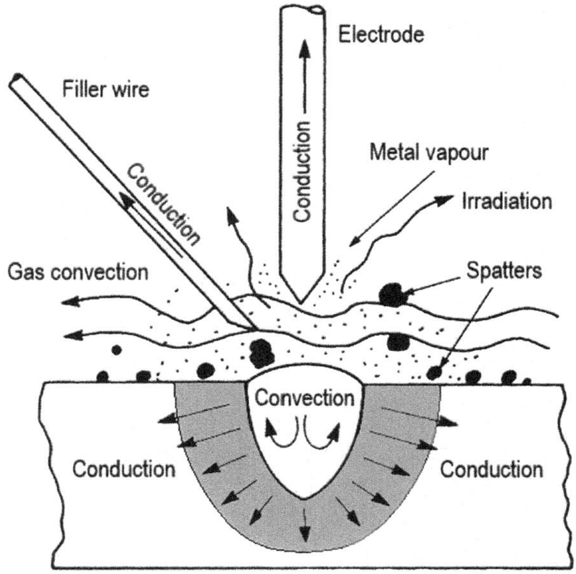

Fig. 3.17 Heat flows for the dissipation of the welding heat input (adapted from Messler, 1999)

3.2.2 Residual Stresses and Solidification Cracking

Weldments are a very popular example for the discussion on thermal stresses. As opposed to mechanical stresses, thermal stresses arise from temperature gradients or temperature changes in a specific spot of a material, resulting in a self-balanced residual stresses field and in distortions. Three main situations can be listed to account for the build-up of thermal stresses: (i) volumetric changes related to thermally-induced phase changes, (ii) mismatch stresses created by temperature changes between two or more materials featuring different values of the coefficient of thermal expansion (CTE), (iii) gradients in expansion or shrinkage of a material subjected to non-homogeneous heating–cooling cycles in its volume.

When the weld metal solidifies, it is subjected to a local volumetric shrinkage which generates thermal stresses considering that the weld pool is always constrained by the surrounding solid. The adjacent regions (HAZ) are also exposed to the spatially and temporally inhomogeneous heating and cooling cycles, with similar effects, albeit with a different timing and kinetics. Considering these thermal stresses, when a welded joint is macroscopically free to accommodate the strains, it will react producing geometrical distortions, as qualitatively depicted in the simplified example of Fig. 3.18. More specifically, when a welded joint experiences constraints at both the macroscopic and microscopic scales, a residual stress field develops during the welding process, as illustrated in Fig. 3.19.

It is expected that a particular material subjected to thermally induced stresses can macroscopically react by producing distortions or, in case of a higher degree of constraint, it can accommodate the local strain field by elastic and plastic deformation.

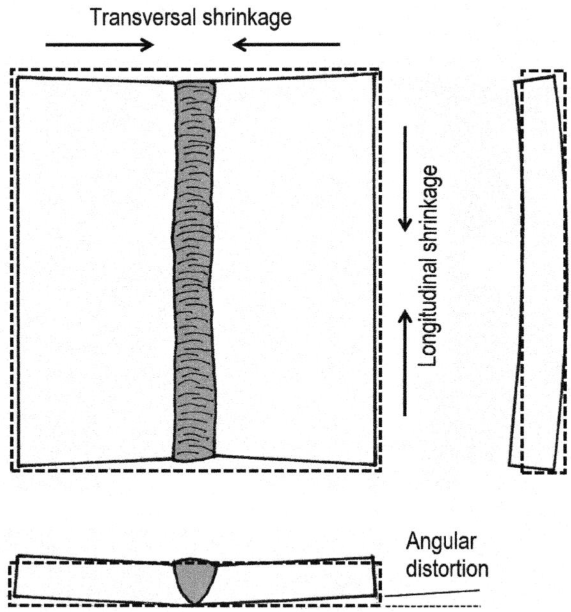

Fig. 3.18 Macroscopic distortions generated in a butt weld joint, broadly classified as transverse, longitudinal and angular distortions

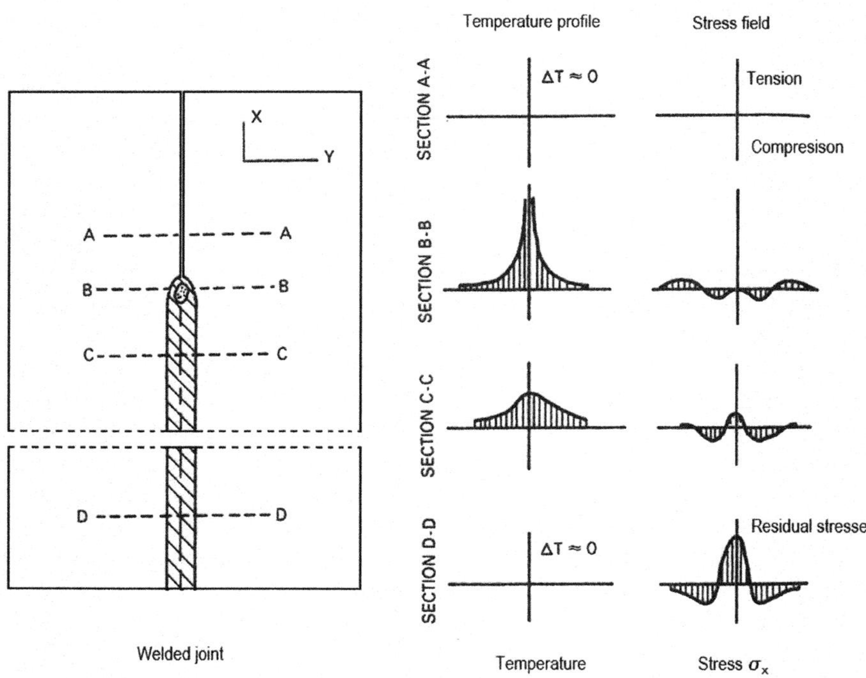

Fig. 3.19 Short-range distribution of residual stresses generated in a butt weld joint (Messler, 1999, published with permission of Wiley Interscience)

Eventually, when the strain exceeds the failure limit of the material, then cracks may appear in the weldment.

Solidification cracking, also called hot-cracking or hot tearing, is a practical evidence of such mechanism. It is occasionally observed in fusion welds in the last stages of solidification, when the accumulated thermal stresses developed across the fraction of already solid grains exceed the failure strength of the weld metal. The most critical scenario occurs when an interlocking network of solid grains is already formed across the two sides of the weld, hence shrinkage stresses can develop, but a semi- continuous film of residual liquid still exists at the grain boundaries, making the material highly susceptible to intergranular separation. Figure 3.20 schematically depicts the described mechanism, which is also common in foundry and ingot castings.

The susceptibility to solidification cracking during welding of an alloy depends on several properties. A fundamental parameter is the solidification temperature range, that rules the width of the interval where both solid and liquid phases coexist, and the amount of liquid phase in the last stages of solidification. Further properties are the strength at high temperature and the hot ductility of the primary solidification phase, the morphology of the grains in the weld metal and the surface tension of the residual liquid at grain boundary, which in turn affects the ability of the liquid to infiltrate creating a dangerous continuous film rather than discrete and disconnected droplets at grain boundaries.

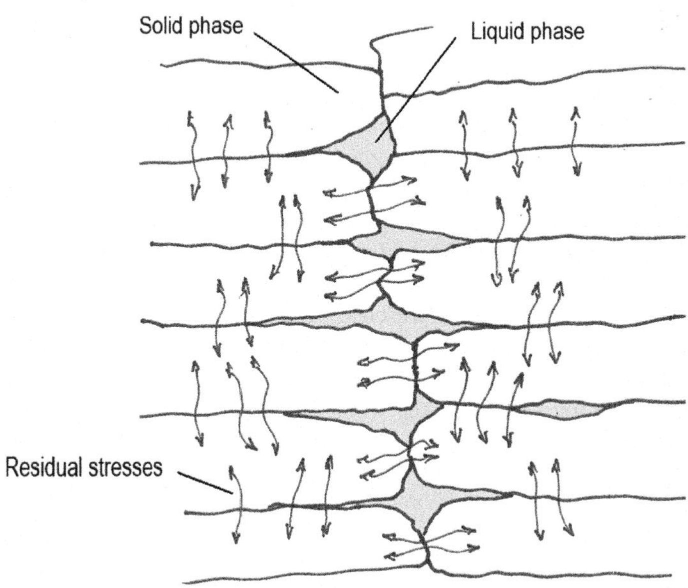

Fig. 3.20 Solidification cracking mechanism developed in the weld metal zone, in the last stages of solidification

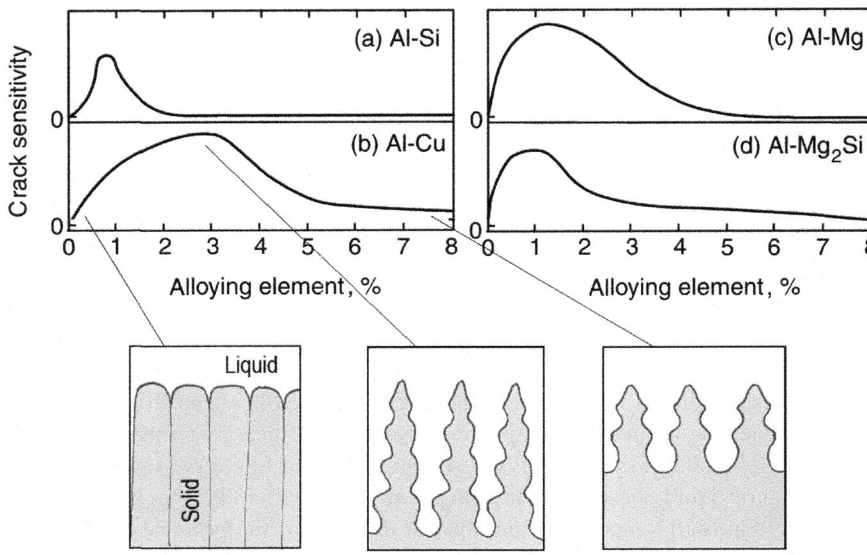

Fig. 3.21 Effect of alloying elements on hot-cracking susceptibility of Al alloy

Concerning the effects of the solidification range and the amount of liquid in the final stages of solidification, it is instructive to consider the data displayed in Fig. 3.21 about the hot cracking sensitivity of some Al alloys depending on their alloying element content (the original data of the graphs have been published by Dudas and Collins, 1966). In general, a wider extent of the freezing range also implies a larger mushy zone where the alloy is weaker due to the co-presence of liquid and solid phases. Figure 3.21 demonstrates that the sensitivity to hot cracking is very low, virtually nil for pure metals, it reaches a peak for intermediate amounts of some alloying elements, and then drops again when the amount of solute increases. The schematics in the figure suggest that for a pure metal, freezing occurs at a fixed temperature, therefore no liquid at solid grain boundaries is expected to exist. At the opposite end, with extensive presence of liquid that promotes the formation of a high fraction of eutectics, any crack that might form is readily backfilled with the available liquid and the crack are healed by the end of the freezing process. The intermediate situation represents the worst case, where the liquid can just form an almost continuous film at the gran boundaries, without any ability to backfill the so-formed cracks.

3.2.3 The Heat Affected Zone in Fusion Welded Joints

The HAZ has been defined as the region of the welded joint that remains at the solid state throughout the welding operation, but it is affected by the welding heat input.

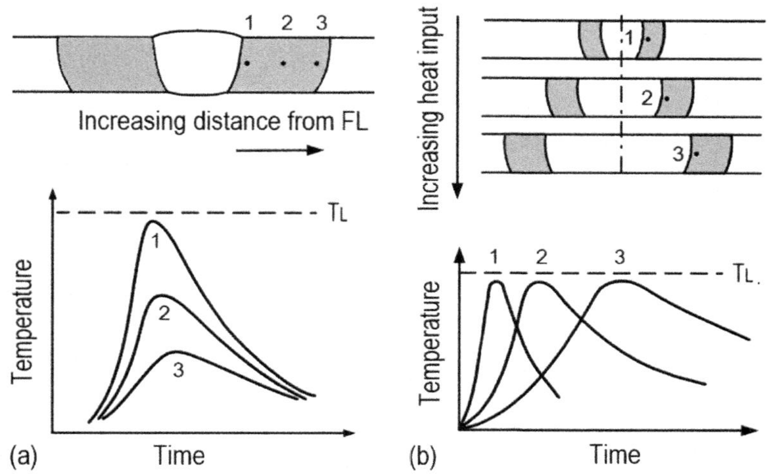

Fig. 3.22 Trends of the welding thermal cycles measured within the HAZ against the distance from the fusion line (**a**) and the welding heat input (**b**) (adapted from Kou, 2003)

The general trend of the welding thermal cycles experienced in the HAZ is depicted in Fig. 3.22a. Increasing the distance from the fusion line (FL) produces smoother thermal cycles featuring relatively slower heating and cooling rates and reaching lower peak temperatures. These features translate into different times spent at high temperature for each specific material portion of the HAZ and in the corresponding expected microstructure modification. The specific position of these thermal curves obviously depends on the net welding heat input and on material thermal properties such as its heat capacity and thermal diffusivity. As for the welding heat input, Fig. 3.22b shows that an increase in the welding power not only generates larger volumes of the weld metal and the HAZ, but also results in longer thermal cycles with expected variation of the microstructure formed, even considering similar distances from the FL.

When discussing about the effect of this exposure to high temperature, it can be observed that the microstructural changes induced in the HAZ will depend on the parent alloy to be welded, namely on its main strengthening mechanisms consisting in one or more of: grain refinement, solid solution strengthening, work-hardening, second-phase precipitate strengthening, transformation (martensite) hardening. Especially sensitive to property changes in the HAZ are the alloys strengthened by work-hardening, second-phase precipitates and martensite hardening.

The effect of the weld heat input on the HAZ of a work-hardened alloy (e.g. an Al alloy belonging to 1xxx, 3xxx, or 5xxx series or a low-C steel sheet) is schematically depicted in Fig. 3.23. Considering the profiles of the thermal welding cycles depicted in Fig. 3.22a, it is suggested that, starting from a defined distance from the fusion line, the exposure to high temperature is such to activate the processes of recovery and recrystallization, replacing the cold-deformed elongated grains with a finer equiaxed structure. Concurrently, the high hardness of the work-hardened parent metal is lost in

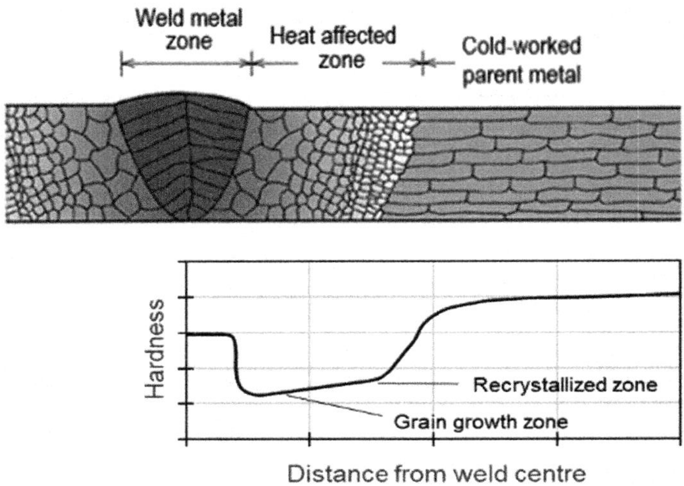

Fig. 3.23 Schematic of the microstructure of a welded joint produced on a cold-worked metal sample and expected profile of the resulting hardness

the HAZ and a clear softening effect appears from the hardness profile. Approaching the fusion line, the higher peak temperature locally reached can also result in grain growth, with a further drop in hardness.

The behaviour of precipitation-strengthened alloys is typically illustrated using as a reference Al alloys belonging to 2xxx, 6xxx, or 7xxx series, γ'-strengthened Ni alloys or precipitation-hardened steels. It is assumed that these alloys are usually treated to achieve their best performance, the peak-hardness aging condition, by a proper solution annealing followed by a fast cooling and an aging treatment to promote the optimum dispersion (in terms of volume fraction and size) of precipitates. Any further heating above the aging temperature after such optimized treatment will deplete or destroy the achieved dispersion of precipitates. As a result, a drop in hardness will become evident due to coarsening of the precipitates at fairly large distance from the FL while the HAZ closer to FL could be subjected to re-solutionizing of most of the precipitates and to the already discussed grain coarsening.

A remedy to the loss of strength in the HAZ of such alloys would be a post-weld treatment that could restore the original precipitate dispersion, when feasible. Alternatively, it can be considered that some Al alloys have been specifically developed to naturally age at room temperature once they achieve a supersaturated condition by a proper solution annealing treatment or by a fairly rapid cooling after reaching a sufficiently high temperature peak, such as in the HAZ of a weld. The graph in Fig. 3.24 provides an example of such behaviour. A 7005 (Al–Mg–Zn) Al alloy plate was welded by a GTAW process using a 5183 (Al–Mg–Mn) Al alloy as a filler metal. Few hours after welding (one day), both the weld metal zone and the HAZ close to FL featured a poor hardness. However, in the HAZ, the 7005 alloy could progressively reach higher hardness values with longer aging times due to natural

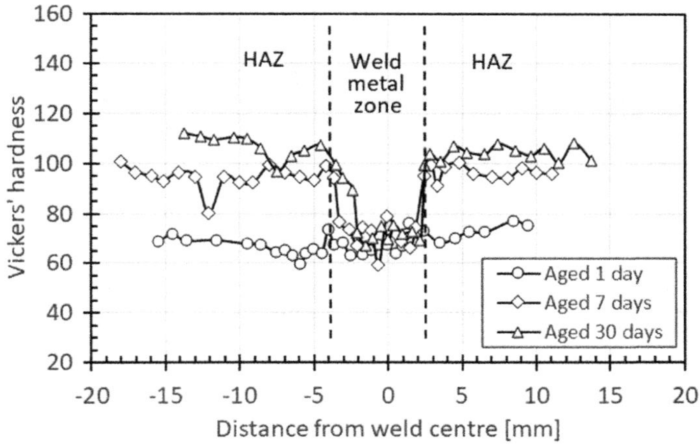

Fig. 3.24 Evolution of hardness with natural aging time across a GTAW joint of a 7005 Al alloy

aging upon storage at room temperature. Conversely, the hardness remained unaltered in the weld metal zone even after one month, due to the modification of the alloy chemistry which was diluted with the filler metal.

Finally, C-steels and low-alloy steels, can be taken as the most popular examples for transformation-hardening alloys that develop martensite upon fast cooling. Martensite forms by a diffusionless, athermal transformation upon cooling from the austenite field at rates exceeding the critical cooling rate for the particular alloy. Under the controlled conditions promoted by a thermal treatment, the steel is deliberately subjected to austenitizing at a temperature where the γ-austenite is stable and then quenched at a controlled rate to achieve martensite. A further stage of tempering must follow to improve toughness of the brittle fresh martensite and relax residual stresses of the quenched part. The ability of a steel to form martensite on cooling (also called hardenability or quench-ability) increases with the amount of C and other alloying elements, but it is also affected by the grain size and homogeneity of the parent austenite, and by possible second phases contained in it.

Hardenable steels are unfortunately prone to a similar uncontrolled transformation also during cooling after welding, resulting in the presence of fresh and brittle martensite in the HAZ of welds. With the combined action of the shrinkage stresses, the presence of undercuts close to weld seam, and the chance to have hydrogen absorbed from the weld metal, cold cracks in the HAZ could easily form, especially in rigidly restrained and thick welded joints. It is to remark that similar concerns should hold in principle also for the weld metal zone where the fast cooling from the liquid phase can induce a similar martensitic transformation. However, a careful selection of the filler metal composition can effectively mitigate this issue by diluting the parent metal and making it less hardenable, hence less prone to produce martensite upon fast cooling.

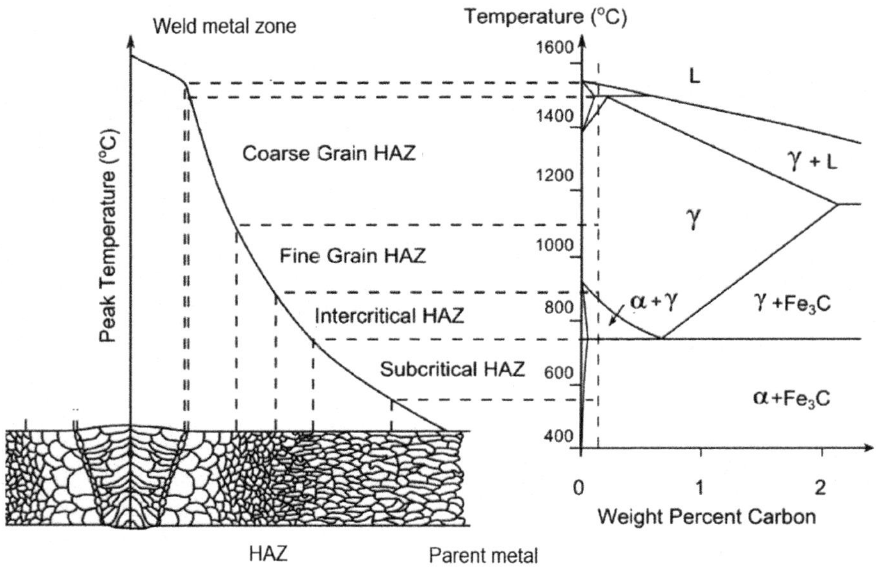

Fig. 3.25 Schematic of the HAZ microstructure developed in a C steel and corresponding phases reached at peak temperature of the welding cycle referred to the equilibrium Fe–C phase diagram

For a given steel, the possibility to develop a martensitic transformation in the HAZ during welding implies the ability to reach, at least partially, the γ-Fe field upon rapid heating and to cool faster than its critical cooling rate. Figure 3.25 shows a schematic representation of the expected microstructure of a C-steel joint and the peak temperature profile experienced during welding. The connected equilibrium Fe–C phase diagram shown on the right side suggests the extension of those regions of the HAZ where the structure could meet the γ-Fe field and potentially lead to transformation into martensite.

It must be specified that the above-proposed scheme brings with it some important approximations. Recalling the thermal welding cycles depicted in Fig. 3.22, it is expected that the high heating rate would modify the positions of the phase boundary lines (hence the critical temperatures denoting the transformation of the steel) in the phase diagram, shifting them to higher temperatures. Moreover, it is also reasonable that the absence of a proper holding time in the welding cycles would leave marked heterogeneities in the high-temperature constituents (e.g. undissolved Fe_3C particles, heterogeneous grain size) that can affect the nucleation of other constituents than martensite upon cooling.

To a first approximation, it is established that the tendency of a steel to develop cold cracks in the HAZ relates to the ability to form martensite, namely on the steel hardenability. For practical purposes, standards like the EN 1011-2 (Welding—Recommendations for welding metallic materials, part 2: Arc welding of steels) suggest simple empirical formulas to evaluate the carbon equivalent (CE) value based on steel composition. For example, for non-alloyed, fine-grained and low-alloyed

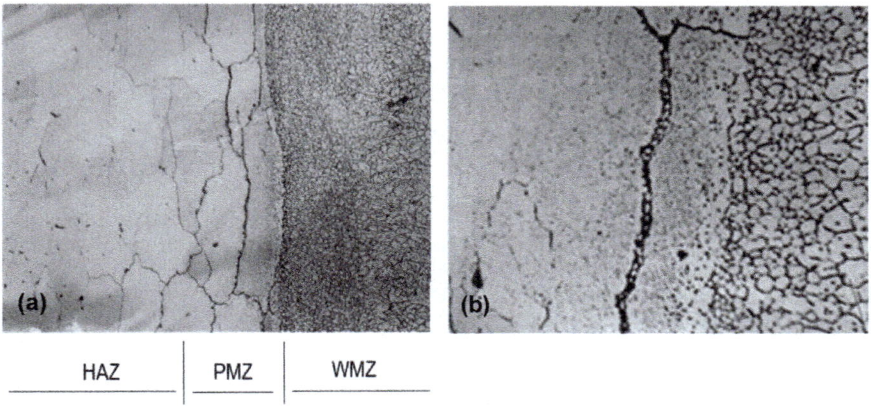

Fig. 3.26 Microstructure of the PMZ formed in a 7050 Al alloy after EBW

steels, the CE (in weight %) is calculated as:

$$CE = C + \frac{Mn + Mo}{10} + \frac{Cr + Cu}{20} + \frac{Ni}{40} \tag{3.4}$$

The higher the CE value, the higher the hardenability and the greater the ability to produce cold cracks in the HAZ. Therefore, it will become more stringent the need to take countermeasures such as increasing the preheating temperature, raising the heat input of the welding process, reducing the potential pickup of hydrogen.

3.2.4 *The Partially Melted Zone*

One last area occasionally found in fusion welded joints of some complex alloys is the partially melted zone (PMZ) which lies immediately outside of the weld metal and faces the HAZ. The material within the PMZ is subjected to partial melting when reaching the peak temperatures of the welding cycles. There, the eutectics or other low-melting constituents generated at grain boundaries due to segregation can reach their melting point promoting the formation of a liquid film, making the PMZ highly susceptible to intergranular cracking (also called liquation cracking in this situation) when subjected to the concurrent shrinkage stresses, according to the mechanisms already described for the weld metal zone.

In Fig. 3.26 the microstructure of a 7050 Al alloy after EBW is depicted. Adjacent to the fine grains formed in the weld metal zone, a thin region corresponding to the thickness of few grains can be identified as the PMZ. Here, thicker grain boundaries are visible and the higher magnification view (see Fig. 3.26b) highlights the cellular structure formed at the grain boundaries after remelting and solidification.

References

Dudas, J.H., Collins, F.R. (1966) Preventing weld cracks in high-strength aluminum alloys. *Welding Journal, 45*, 241

Kou, S. (2003). *Welding metallurgy* (2nd ed.). Wiley.

Messler, R. W. (1999). *Principles of welding—processes, physics, chemistry and metallurgy*. Wiley.

Messler, R.W. (2011). ASM handbook. In T. Lienert, T. Siewert, S. Babu, V. Acoff (Eds.), *Welding fundamentals and processes—overview of welding processes* (Vol. 6A). ASM International.

Piatti, G., & Vedani, M. (1989). Material variability in composition and structure of a type 316 stainless steel submerged arc weld metal. *Journal of Materials Science, 24*, 1429–1438.

Roberts, P. (2013). *Industrial brazing practice*. CRC Press Publisher.

Way, M., Willingham, J., & Goodall, R. (2020). Brazing filler metals. *International Materials Reviews, 65*, 257–285.

Chapter 4
Metal Powders and Powder Metallurgy

Metal powders are used in a wide number of applications as feedstock material to build functional and structural parts such as filters, heat exchangers, self-lubricating bearings, small tools and tool inserts, gears and other precision mechanical components. The consolidation of the powder into the final shape involves the densification of the powder particles at the solid or semi-solid stage, as in the case of press and sintering, or at the liquid stage, typically in some additive manufacturing processes.

The synthesis of metal powder by itself involves dedicated processes based on the atomization of a liquid melt into droplets, or on the mechanical crushing and milling of solid metal fragments. Alternatively, physical–chemical methods based on electrolytic or chemical routes can be used for the production of small batches of powder. Powders, with their peculiar state of aggregation, also feature specific properties and require dedicated characterization methods to quantify their size and shape, their apparent density and flowability, just to mention few.

In this chapter, the main powder production processes are preliminary described and the different powder properties achievable depending on the manufacturing route are discussed. The focus will then be addressed to the fundamental principles of powder metallurgy, to elucidate the phenomena involved in the densification of powders using well established industrial processes based on solid-state (or semi-solid state) sintering. The subject of powder consolidation by additive manufacturing methods will instead be covered in a distinct chapter, due to the growing importance of such innovative processing route.

M. Vedani, *Metal Science in Modern Manufacturing Technologies*, Springer Series in Materials Science 351, https://doi.org/10.1007/978-3-031-97125-9_4

4.1 Powder Manufacturing

4.1.1 Atomization Processes

The most common and cost-effective method for the production of low-reactive metal powders (e.g. iron and steels, nickel alloys, copper and bronze) is water atomization (WA) of a liquid melt. As depicted in the schematic image of Fig. 4.1, WA is a batch process whereby an overheated molten alloy is poured into a tundish. A nozzle on its bottom allows a fine melt stream to fall in the atomization chamber where it is disintegrated into small droplets by a set of high-pressure water sprays. The so-formed tiny droplets rapidly solidify while falling along the chamber and are collected on the bottom as solid powder particles mixed with water. The mixture is then extracted from the chamber, dehydrated, sieved according to the desired size range, annealed and packed into sealed containers.

If the WA route covers the lion's share among the atomization processes in terms of production tonnage, gas atomization (GA) processes raised the attention of metal powder industries for the synthesis of more reactive alloys and for higher quality of powders dedicated to specific processes, In GA the molten metal stream is atomized by using a high-pressure flow of inert gas, typically argon or nitrogen, instead of water. The relatively lower cooling rate induced by the gas leaves the surface tension a longer time during the flight within the atomization chamber, generating droplets with a nearly spherical shape, which is particularly suitable for powder-bed based additive manufacturing processes (see Chap. 5), as opposed to the irregular morphology created by WA. In addition, the inert environment in the chamber protects the metal

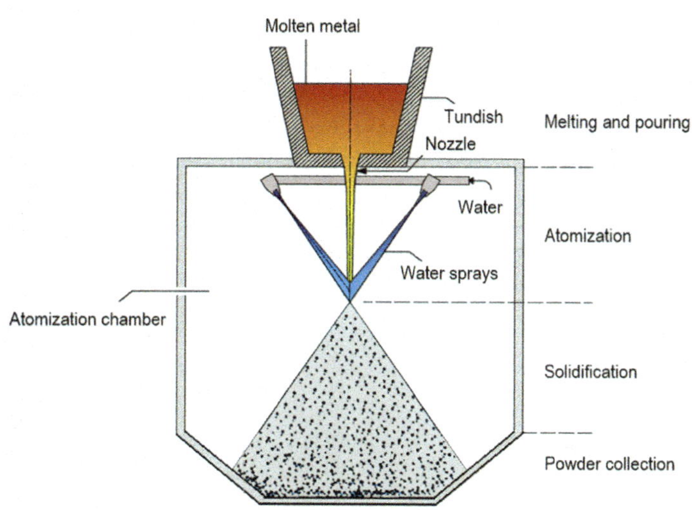

Fig. 4.1 Schematic of the water atomization process (adapted from Asgarian et al., 2020)

from oxidation, making the GA process suitable for the production of high-purity powders, even with reactive metals.

Figure 4.2 shows some of the main existing variants of GA processes. Vacuum induction melting atomization (VIGA) relies on the induction melting of the metal charge in a crucible placed in a protected chamber above the atomizer, as shown in Fig. 4.2a. For high-melting point and more reactive metals, the electrode induction melting atomization (EIGA) route shown in Fig. 4.2b is preferred since any contamination of the alloy with the crucible is prevented by the non-contact direct melting of the feedstock bar within an induction coil. However, the main limitation is given by the need of high-quality alloy bars as the starting electrodes, which restricts the material choice and raises the cost.

The plasma rotating electrode process (PREP) shown in Fig. 4.2c involves spinning a bar of the material to be atomized at high rotational speed while its end is melted by a plasma torch. The centrifugal force pulls the liquid radially outwards and generates droplets that solidify in the chamber to produce perfectly spherical and clean powder particles. An alternative process using the same heat source is the plasma atomization of wires. depicted in Fig. 4.2d. Here, plasma torches are used to melt the tip of a wire continuously fed in the atomization chamber. This process has been developed for the production of high-purity powders of reactive metals like titanium, tantalum, zirconium, with fine size and good sphericity. An obvious disadvantage is that the feedstock alloy needs to be pre-processed into a wire, which prevents the use of the most brittle and hard alloys,

Finally, a new emerging route for metal powder production is represented by ultrasonic (US) atomization, as reported in Fig. 4.3. The process consists in contacting the liquid melt to the surface of an US transducer, the sonotrode, promoting the fragmentation of the liquid volume into fine droplets and leading to a final powdered solid product with high sphericity.

The precise mechanisms governing the drop formation in a liquid shaked within the US frequency is still a matter of scientific debate. In many publications two fundamental hypotheses are raised, with the suggestion that the actual mechanism might be a combination of them, strongly dependent on the density, viscosity, surface tension and temperature of the specific alloy being processed. According to the first hypothesis, capillary waves are formed on the surface of a liquid once it is subjected to continuous oscillations induced by the sonotrode (with frequency in the approximate range 20–100 kHz). These waves are made by peaks and valleys and once their amplitude exceeds a critical threshold, some peaks can separate from the surface of the waves overcoming the surface tension (see Fig. 4.3). Small droplets are thus generated and are ejected out of the liquid volume. In a second hypothesis, cavitation is considered as the triggering effect for droplet formation. Under the action of the US oscillations, cavitation is created within the liquid volume where the pressure drops below its vapor pressure. Tiny vapor bubbles can rapidly grow and collapse under the fluctuating pressure, thus generating micro-shock waves which can reach the surface and expel droplets from the liquid volume. As, already mentioned, a reasonable idea is that a synergistic combination between the two above hypotheses could take place,

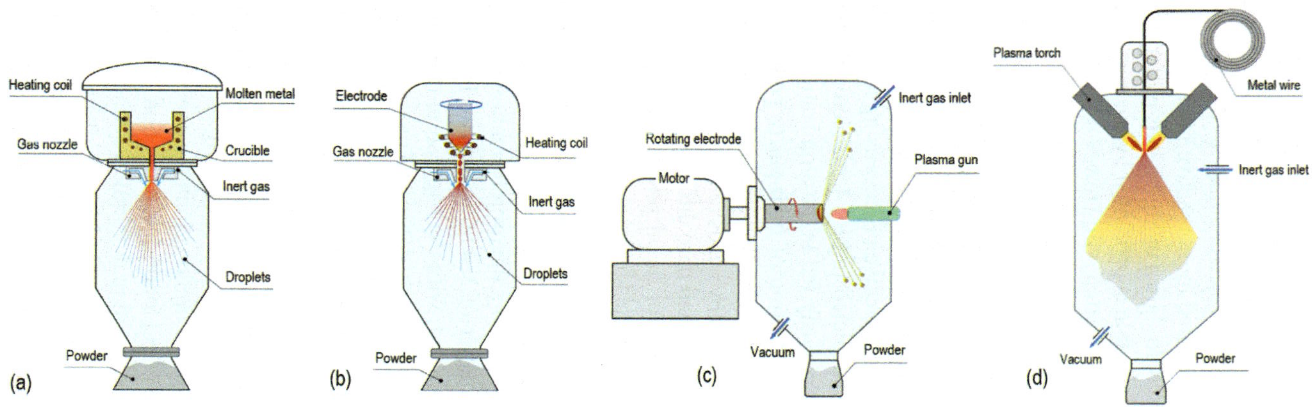

Fig. 4.2 Schematics of the main gas atomization processes. **(a)** Vacuum induction melting atomization, **(b)** electrode induction melting atomization, **(c)** plasma rotating electrode process, **(d)** plasma atomization (Feng et al., 2024, published with permission of Springer Nature)

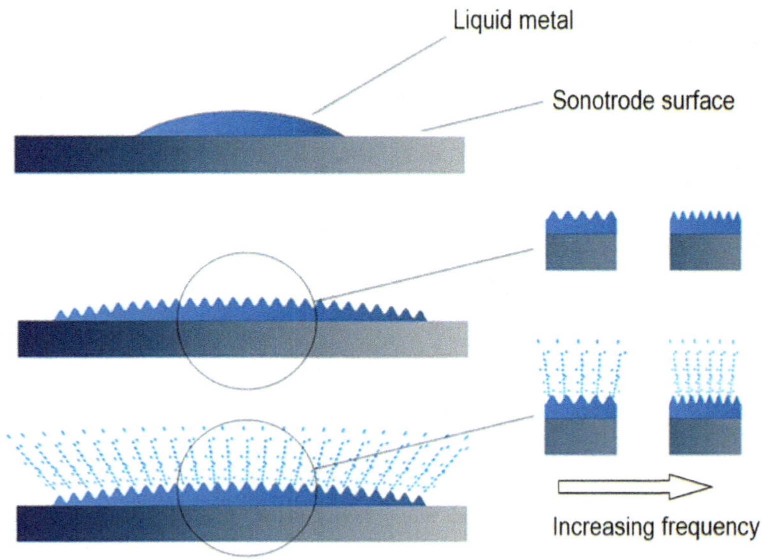

Fig. 4.3 A simplified representation of the liquid metal droplet generation by ultrasound atomization (adapted from Kustron et al., 2023)

whereby cavitation generates high-frequency shocks that stimulate capillary surface waves, ultimately causing droplets to be ejected from liquid surface.

For practical applications in powder production, the sonotrode must be coupled to a heat source enclosed within a sealed chamber, capable of melting the alloy and maintaining it at the appropriate temperature for the atomization. Figure 4.4 shows a schematic setup of a possible US atomization system. The heat can be provided either by a plasma torch or a laser source, while the feedstock alloy can be delivered as a filler wire or as a consumable electrode. The process preferably lend itself for small batch production of powder and for the testing of experimental alloys.

4.1.2 Mechanical Milling Processes

Mechanical methods involve loads such as those generated by compressive, shear or impact forces to facilitate particle size reduction of bulk materials. Terms like milling, crushing, grinding, and comminution, are often used to identify these processes in different industrial fields and for specific purposes.

The general milling mechanisms is based in the idea that the application of an external force to a block of material generates stresses leading to deformation (elastic and plastic) and eventually cracking, provided the material is sufficiently brittle. The original blocks are then progressively crushed into small fragments till the

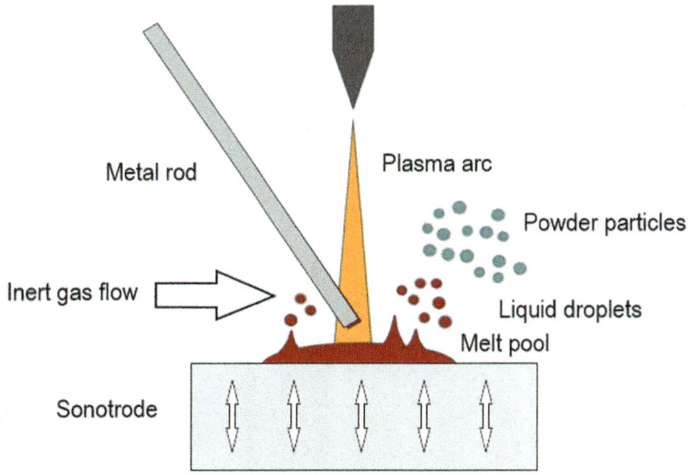

Fig. 4.4 Schematic setup of an ultrasonic atomization process

desired powder size is achieved. In principle we can distinguish two types of crushing processes, surface-based and volume-based crushing, as represented in Fig. 4.5.

For metal powders with a more ductile behaviour, the reduction of particle size and the modification of the powder itself is often achieved using ball milling (BM) techniques. There exist several variants of the BM process as summarized in Fig. 4.6a. The common principle is that feedstock metal parts or coarse powder are filled in rotating drums that also contains a significant fraction of hard and heavier balls (made of ceramic, stainless steel or other inert material) and in some cases a process-control agent to limit sticking of particles during processing. Upon operation of the BM equipment, the powder particles are trapped within the colliding balls and undergo extensive mechanical loads at fairly high strain rates. The resulting strain energy leads to a sequence of transformations summarized in Fig. 4.6b, eventually refining and homogenizing the powder.

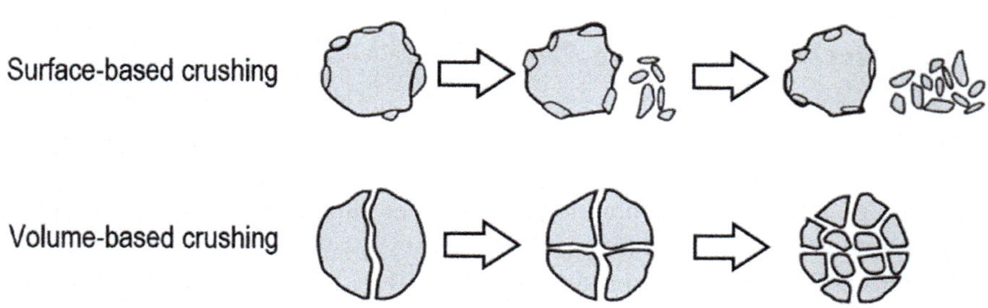

Fig. 4.5 Simplified mechanisms of surface and volume crushing in powder manufacturing

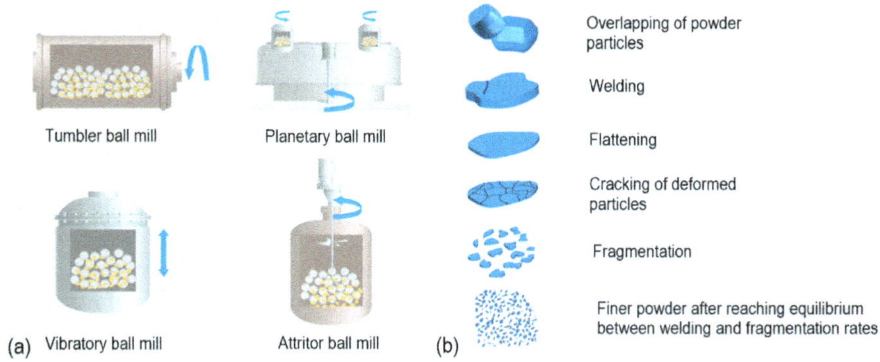

Fig. 4.6 Principles of ball milling for metal powder production. (**a**) Schematics of the milling systems, (**b**) mechanisms of powder size refinement (Dhiman et al., 2021, published with permission of Elsevier)

4.1.3 Other Processes

For sake of completeness, some other powder production processes in use for metals will be briefly mentioned in this section. Solid state reduction of ores (particularly oxides) using solid or gaseous reducing agents is a common industrial method to produce metal powders such as Fe, Cu, Ni. It has been for a long time the most widely used method for the production of iron powder. Selected ore rich in Fe is preliminarily crushed and blended with carbon before being treated in continuous furnaces where it is reduced to a porous sponge of metal. The sponge is then separated from residual oxides, milled, and sieved to the desired powder size.

Electrolytic deposition is a further process to produce high-purity metals at the cathode of an electrolytic cell filled with aqueous solutions or fused salts. In general, the deposit generated at the cathode can be hard and brittle, which is subsequently milled to obtain powder, can be spongy and loosely adherent, which is easily removed by scrubbing, or could even be a deposit already in powder form, that collects at the bottom of the cell.

4.2 Powder Properties and Their Characterization

Powders are characterized by their peculiar physical and technological properties which are in turn related to the expected processing and application requirements. The specifications for powder raw materials imply the definition of somewhat unconventional properties when compared to bulk metals. Among others, powder composition (with special care to surface contaminants), particle-size distribution, powder morphology, density, flowability are some of the first parameters with direct effect on final part performance after powder consolidation (Neikov et al., 2009).

To start with, powder chemical composition is usually measured with the same methods also adopted for bulk materials. However, due to the remarkably higher surface-to-volume ratio of fine powders, the monitoring of interstitial elements becomes particularly important, especially for the most reactive metals and alloys. Pickup of oxygen is easily detected in powders exposed to air environment and after repeated recycling of the powder. Humidity can also be promptly adsorbed on powder particle surfaces, resulting in increased amount of oxygen and hydrogen in the final product.

Depending on powder consolidation process, there exist most appropriate ranges of particle size, spanning from few micrometers up to several hundreds of micrometers. The particle size distribution (PSD) is therefore a variable of paramount importance for metal powders. It can be measured by sieve analysis, where cloth sieves woven in a square mesh pattern with increasing size are used to split the particle population into several size groups and to separate particles outside of the allowable range. More frequently, laser diffraction methods are adopted. This technique is relatively fast and ensures the accurate measuring of particles as small as 0.1 μm. It is based on the scattering of the laser light by the dispersion of the powder, considering that each particle in a parallel laser beam deflects the light at a fixed angle that depends on the particle diameter.

A typical output of the particle size analysis is a graph reporting the PSD as the frequency distribution of the particles against their size, as shown in Fig. 4.7. In simplified terms, the PSD can also be reported using the D_{10}, D_{50} and D_{90} percentiles, which represent the size of the particles covering the lower end at 10%, 50% and 90% of the cumulative curve, respectively.

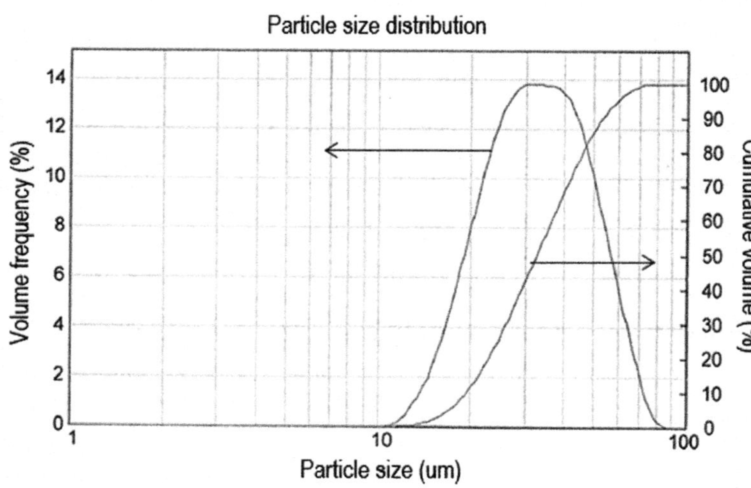

Fig. 4.7 PSD plot for metal powder showing the volume frequency and cumulative volume versus the particle size

The shape of the powder particles represents a further variable to be carefully considered. Powder morphology can be evaluated by optical microscopy, but more representative results can be gathered using scanning electron microscopy (SEM). Figure 4.8 reports a classification of the various particle shapes based on their dimensionality (Nouri & Sola, 2018).

Indeed, in previous sections it has been highlighted that, depending on the powder production process, not only the size, but also the shape can significantly differ. The particles are almost perfectly spherical in gas atomized powders, more irregular in water atomized powders and markedly sharp-edged in crushed powders. Porous or dendritic, and more irregular shapes are often found in powders obtained from reduced ores and electrolytic processes. Figure 4.9 shows some representative SEM images of powder particles.

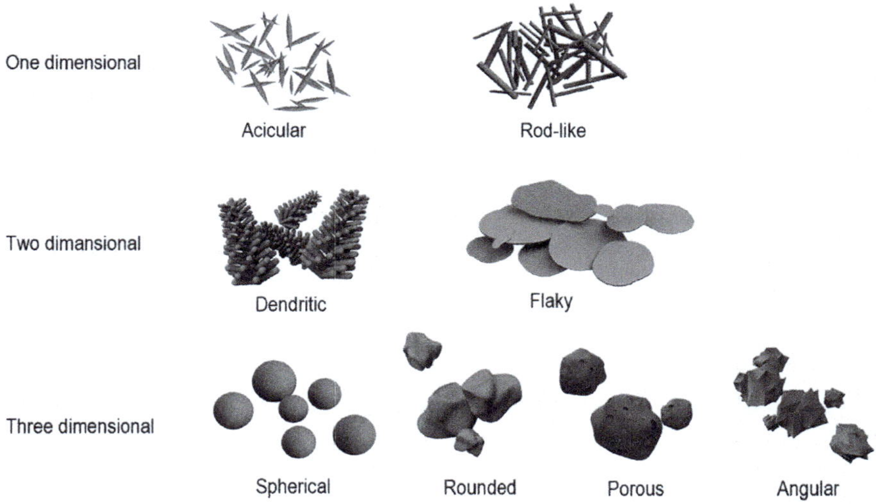

Fig. 4.8 Classification of the particle shapes based on their dimensionality (adapted from Nouri & Sola, 2018)

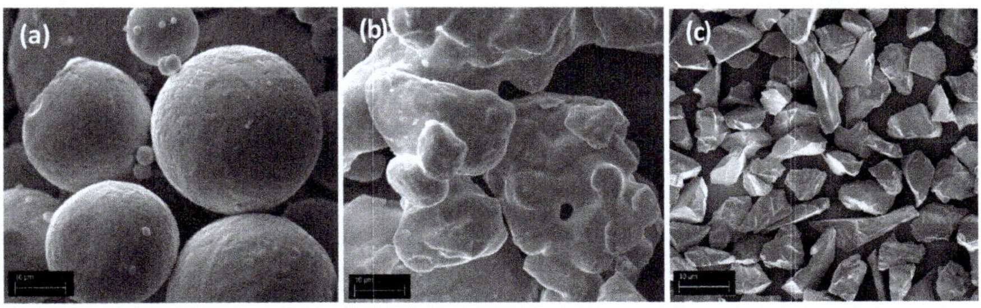

Fig. 4.9 Powder particle morphology observed by SEM images on (**a**) gas atomized steel, (**b**) water atomized steel, and (**c**) crushed WC

Additional information may refer to the bulk properties of a given powder volume and its flow behaviour during die filling or powder bed spreading. The apparent density is a first measure of the weight of a unit volume of loose powder once it flows freely (ruled by gravity) through a funnel in a density cup. Specific standards exist for the correct measurement of the apparent density by this so-called funnel method, such as the ISO 3923-1 2018 (Metallic powders—Determination of apparent density Part 1: Funnel method), and the ASTM B212 2021 (Standard test method for apparent density of free-flowing metal powders using the Hall flowmeter funnel). For a given metal, the apparent density generally decreases with decreasing particle size and as the shape becomes more irregular. A wider PSD contributes to increase the apparent density because the spaces between the coarser particles can be readily filled with the smaller ones. The Packing density (also called tapped density) refers to the powder density once the volume filled in the density cup is gently tapped or vibrated under specific conditions. Tapping allows the rearrangement of the powder particles and the reduction of the occupied volume.

In Table 4.1 representative data about the densities measured for some steel powders are reported. It is demonstrated that the apparent density of a metal powder roughly corresponds to 50% of the material bulk density and that tapping can promote a densification of about 20% with respect of the free-flowing powder.

Finally, quantification of the ability of a powder to rapidly and homogeneously flow when filling a cavity or when generating a uniform layer in powder-bed based additive manufacturing is of great importance in industrial processes. A first information is collected measuring the flow rate, as the time required for a standard sample (usually 50 g of powder) to flow through a standard Hall funnel into a cup. A second approach considers the angle of repose of a cone of powder freely poured through a funnel onto a flat surface. The angle of repose α is defined as the angle of the so-formed powder cone with the horizontal plane, as shown in Fig. 4.10a. It is related to the interparticle friction, ultimately the flowability, under static conditions. In an analogous way, dynamic flowability is evaluated filling half of a drum with transparent sides with powder and evaluating the dynamic angle assumed by the powder while the drum is rotating at a moderate speed (see Fig. 4.10b).

Table 4.1 Comparison of densities of four gas-atomized steel powders (data from Yan et al., 2019)

Material	Apparent density (g/cm^3)	Tapped density (g/cm^3)	Tapped/apparent density ratio	Bulk density (g/cm^3)
316L	4.27	5.12	1.20	8.00
H13	3.95	4.94	1.25	7.80
P20	4.21	5.03	1.19	7.85
18Ni300	4.19	5.03	1.20	8.00

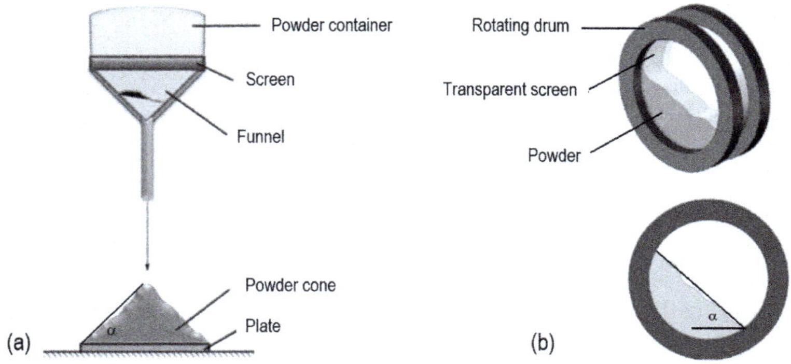

Fig. 4.10 Definition of the testing setup and of the angle of repose α for the evaluation of static (**a**) and dynamic (**b**) flowability of powders (adapted from Muthuswamy, 2022)

4.3 Powder Metallurgy Processes

Powder metallurgy (PM) is the term referred to the manufacturing of solid parts of desired geometry starting from powders and exploiting a sintering mechanism. Since similar processes are also used for the consolidation of non-metal powders such as ceramics and hybrid powder blends, the general term of powder sintering is also frequently used.

PM lend itself to the production of cheap, complex-shaped mechanical parts, porous materials such as filters and oil-retaining bearings, components with refractory metals that are otherwise difficult to be produced by casting or plastic deformation, tools made of sintered carbides with Co or Ni matrix and diamond cutting tools.

As summarized in the workflow of Fig. 4.11, after powder preparation and tooling, the process involves a first stage of pressing the powder in a die with the desired shape to obtain a "green" compact, strong enough to be extracted from the die, gently handled and transferred to a sintering furnace where, in a protected atmosphere, it is heated to achieve the solid-state welding (i.e. the sintering) of the individual powder

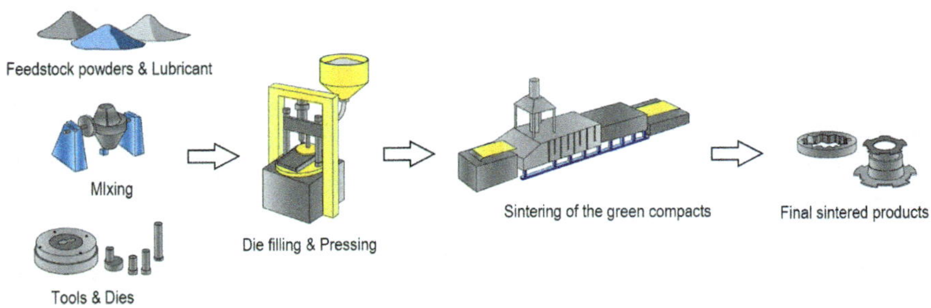

Fig. 4.11 The general workflow of the press and sintering process

particles. Further details about this process of "press and sintering", together with few possible variants of the traditional route, will be described in the next sections.

4.3.1 Powder Blend Preparation and Pressing

The first operation in view of press and sintering consists of the feedstock powder preparation. The alloy powder with the desired composition (if it is not the case of pure metal parts) shall be created according to one of the methods depicted in Fig. 4.12. The use of pre-alloys powders, when available, is always preferred, but alternative routes could be mixing different elemental powders (provided their melting point is not too much different), correct powder composition by surface diffusion or surface modification, or using a combination of these strategies. When working with heterogeneous powder particles it is implicitly assumed that the following high-temperature sintering stage should be able to promote enough elemental diffusion to achieve a homogeneous composition in the final sintered part.

Additional lubricant can be incorporated within the powder to reduce the friction between the particles and the die walls, and to a limited extent also among the powder particles, thus improving the following pressing operation along with the safe extraction of the green compact from the die. Popular lubricants are stearic acid, zinc stearate and other organic compounds, added to the feedstock in a fraction of up to 1% by weight.

The powder blend is then poured into the die cavity and pressed to shape usually at room temperature by promoting the mechanical interlocking of the powder particles under the action of a pressure within the range 150–1400 MPa, depending on material, part thickness and expected performance. A relevant issue in this phase is that the pressure within the powder volume cannot be transmitted uniformly as it would happen in a liquid. Indeed, friction progressively reduces the pressing action with increasing distance from the moving surface of the die (the punch). Figure 4.13 qualitatively depicts the pressure distribution below the punch during a pressing operation. To achieve the proper pressure also at distance from punch surface, one should increase the applied load. However, there are limitations related to tool failure (which also increases with complexity of the die), punch buckling and press capacity.

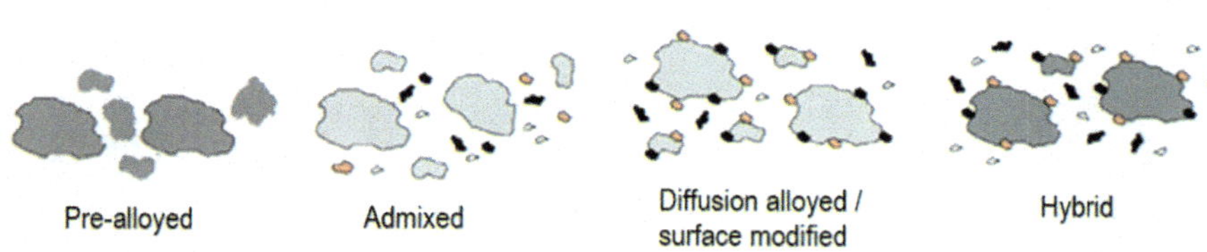

Fig. 4.12 Different strategies to create alloys with metal powders

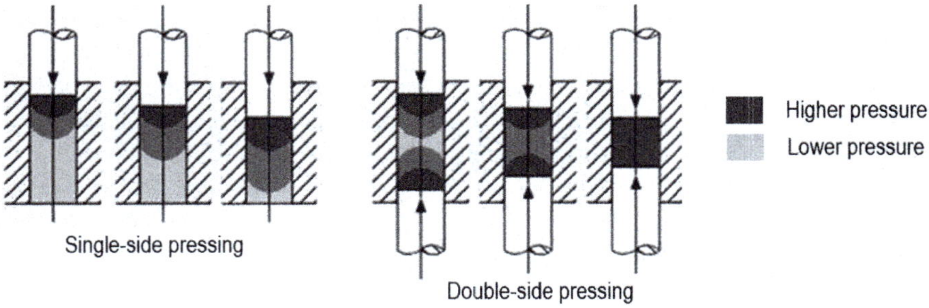

Fig. 4.13 Powder pressure distribution achieved within a die by single-action and double action pressing

An effective countermeasure to reduce the pressure drop is the design of double-action tools which can better distribute the pressure by using two or more punches operating from above and below the die.

The pressure to be reached at any position of the green part volume depends on the desired amount of cold compaction, evaluated by measuring the green density. It is based to a large extent on material strength but is also affected by powder characteristics (e.g. powder morphology, PSD) and by the effects of the lubricant. The compaction behaviour of a specific powder blend is experimentally evaluated by measuring the achieved relative density (i.e. the actual density of the compact divided by the bulk density of the material) against the amount of applied pressure. These data are gathered in typical compaction curves, as shown in Fig. 4.14.

The starting point of the compaction curve corresponds to the apparent density of the powder blend achieved once it fills by gravity the die. As the compression starts,

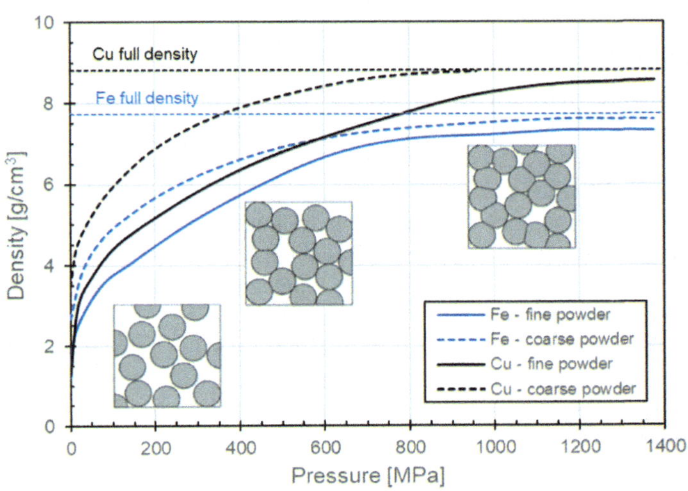

Fig. 4.14 Schematic trends of compaction curves for different powders

the volume of the powder decreases by the rearranging of the individual powder particles, resulting in an increase of stable contact points. Increasing the pressure leads to the onset of a plastic deformation stage where the contact areas are extended and mechanical interlocking at particle surfaces is created. Further material deformation is increasingly opposed by work-hardening and by friction, producing a saturation of the compaction curve (i.e. the density increase becomes smaller with increasing applied pressure). The density reached at end of the pressing stage, corresponding to the green density, is always somewhat lower than the material theoretical density due to the presence of a network of interconnected residual pores and the possible presence of the lighter lubricant.

4.3.2 Powder Sintering

Sintering is the core stage of the PM process, where the green compact is held at high temperature to acquire a higher strength and increase its density, shifting from the mechanical links existing among powder particles of the green part to the wider and stronger diffusion-welded areas of the sintered part. Sintering is carried out most frequently in continuous furnaces, at temperatures between 60 and 90% of the melting-point of the main alloy constituent. For some powder mixtures, sintering may even take place at a temperature exceeding the melting point of the lower-melting constituent. The presence of a liquid phase significantly accelerates the process of mass transfer, but the amount of liquid shall be limited and well controlled to avoid impairing the geometrical shape of the parts.

The selection of the sintering atmosphere of the furnace is of crucial importance. The right atmosphere should prevent powder oxidation, or even promote the removal of the existing oxides by chemical reduction, it should avoid other unwanted reactions like decarburization and assist the burning of the lubricant and any other additives. Common atmospheres used in industrial processes are dry hydrogen, nitrogen/hydrogen mixtures and partially combusted hydrocarbons. Sometimes, sintering may be performed in inert gases or in vacuum for some refractory metals.

During the first stages of sintering, while raising the temperature, the lubricant is first burned off. Once the sintering temperature has been reached, the extensive diffusion promotes the conversion of the individual particle structure into a continuous microstructure by virtue of several mechanisms such as extensive plastic flow, recrystallization, grain growth, liquid-phase and vapour-phase mass transport. These processes are summarized in Fig. 4.15.

During sintering the porosity is also reduced, moving from the interconnected network of pores into a closed porosity made of fewer isolated pores. It is to remark that the reduction in porosity, hence the increase in density for the same mass, invariably results in a shrinkage of the sintered part. This is of limited concern in controlled industrial processes, since the shrinkage can be compensated at the design stage by increasing the green part size.

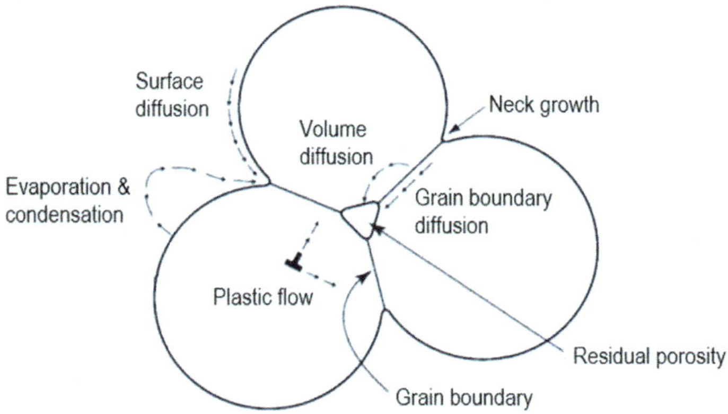

Fig. 4.15 The diffusion mechanisms and microstructure development acting during sintering

Following the sintering stage, the parts generally feature all the required properties for the final application. However, for specific requirements and materials, further finishing operation can be adopted. Infiltration consists of filling the pores exposed on the surface with a liquid metal having a lower melting point. Typically, sintered steel parts are infiltrated with copper. Impregnation is a similar process except that the filler is an organic material instead of a metal, the most popular example being oil-impregnated bearings. Steam treatment is a further process peculiar for PM steel parts. It involves the exposure of the parts at around 500 °C to high-pressure steam leading to the formation of an oxide surface layer which improves hardness, corrosion and wear resistance, reduces surface porosity and seals possible interconnected pores to assure tightness of the PM parts.

Heat treatments can be applied to suitable alloys to increase strength, even though this is not the rule for PM parts. Finally, machining might be required for fine details (e.g. threads and lateral holes) in some PM parts, while sizing can be used to improve geometrical tolerances, and coining could be beneficial for surface finishing and to add small details on surfaces.

4.3.3 Alternative PM Processes

Even though the above-described press and sintering is by far the most popular PM process, other variants for the densification of metal powder exist. Their main principles will be briefly described in this section.

Hot pressing involves the pressing and sintering operations of a part in a single step, as shown in Fig. 4.16a. Since at high temperature metals are softer, it is possible to press the part reaching much higher density with reduced load, resulting in good mechanical properties and enhanced dimensional accuracy. The main limitations lie in the higher cost of the tooling, which should be equipped with heaters and be made

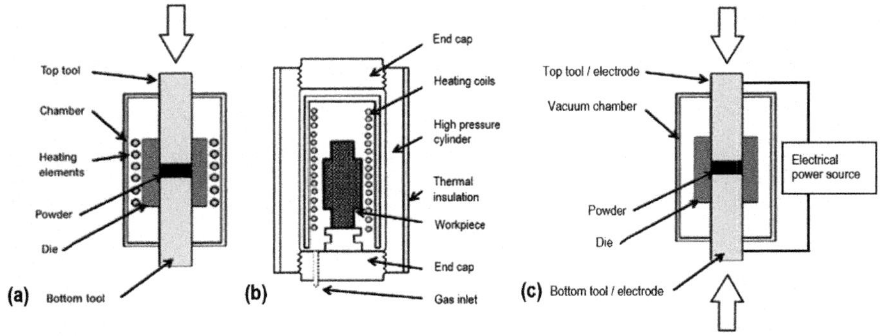

Fig. 4.16 Advanced PM processes derived from the traditional press and sintering; (**a**) hot pressing, (**b**) hot isostatic pressing, (**c**) spark plasma sintering

of heat resistant materials. Moreover, the longer processing times, which include the heating and the cooling stages at each cycle, and the need to provide a protective environment for the powder held at high temperature make the processing cycles longer and more complex.

Isostatic pressing is a further approach whereby the usual unidirectional pressing is replaced by the action of a fluid which exerts the pressure from all directions. In cold isostatic pressing (CIP) the powder is sealed in a flexible container made of an elastomer which sets the shape of the part and it is then immersed in water or oil pressurized at 400–1000 MPa. Due to the radial action of the pressure, the green compact achieves a uniform density even without the need of any lubricant. A conventional sintering stage will follow to reach full strength of the part. A more used variant of the above process is represented by the hot isostatic pressing (HIP). In HIP (see Fig. 4.16b) the powder is fed in a metal sheet container reproducing the simplified shape of the desired part and the fluid used to pressurize the container is an inert gas like argon. The pressure vessel with the powder container is then held at high temperature at a pressure up to 200 MPa. The concurrent application of the isostatic pressure and of the heat allows achieving parts with almost full density and excellent mechanical properties. The tooling cost, cycle times and energy requirements are obviously much more demanding than those required for the traditional press and sintering.

It is worth considering that HIP is also exploited as a secondary process to reduce porosity of parts already manufactured by other routes. A common practice consists of HIP-processing critical parts produced by castings, additively manufacturing, already sintered cemented carbides and special alloys, to eliminate any residual defect and increase reliability in service.

Finally, Spark Plasma Sintering (SPS) is a sintering technique that involves the concurrent application of a low-voltage high-density pulsed current and the uniaxial pressure to the powder within a graphite die, as schematically depicted in Fig. 4.16c. According to established theories, the short current pulses crossing the powder generate a rapid Joule heating at the contact points of the particles and spark discharges with occasional production of a plasma at non-contacting particle gaps. This process promotes the disruption of the oxide layer and rapidly activates the sintering process. The sintering is assisted by the local melting and evaporation at the surface of the particles and the plastic deformation of the volume immediately beneath the surface due to the applied pressure. The SPS process typically takes place in a vacuum or controlled atmosphere, within a graphite die and punches that also act as electrodes. The whole SPS process requires a relatively short time (in the range of few dozens of minutes) and can be carried out at lower temperatures than in conventional press and sintering. Moreover, since most of the heat is localized at particle surfaces, there is a stricter control on bulk microstructure, especially in terms of grain size. SPS has found applications in a variety of fields to fabricate part with simple geometrical shape made of metallic, ceramic, and composite materials with high density, fine grain size, and enhanced properties (Cavaliere, 2019; Mamedov, 2002).

References

Asgarian, A., Tanga, Z., Bussmann, M., & Chattopadhyay, K. (2020). Water atomisation of metal powders: Effect of water spray configuration. *Powder Metallurgy, 63*(4), 288–299.

Cavaliere, P. (2019) Spark plasma sintering: Process fundamentals. In P. Cavaliere (Ed.), *Spark plasma sintering of materials: advances in processing and applications*. Springer Nature. https://doi.org/10.1007/978-3-030-05327-7_1

Dhiman, S., Joshi, R. S., Singh, S., Gill, S. S., Singh, H., Kumar, R., & Kumar, V. (2021). A framework for effective and clean conversion of machining waste into metal powder feedstock for additive manufacturing. *Cleaner Engineering and Technology, 4*, Article 100151.

Feng, S., Jia, D., Fu, Y., Kong, X., Lv, Z., Zeng, E., & Gao, Q. (2024). Preparation of additive manufacturing powder by external field–enabled: A comparative assessment. *The International Journal of Advanced Manufacturing Technology, 131*, 3239–3265.

Kustron, P., Korzeniowski, M., Sajbura, A., Piwowarczyk, T., Kaczynski, P., & Sokolowski, P. (2023). Development of high-power ultrasonic system dedicated to metal powder atomization. *Applied Sciences, 13*, 8984.

Muthuswamy, P. (2022). Infuence of powder characteristics on properties of parts manufactured by metal additive manufacturing. *Lasers in Manufacturing and Materials Processing, 9*, 312–337.

Mamedov, V. (2002). Spark plasma sintering as advanced PM sintering method. *Powder Metallurgy, 45*(4), 322–328.

Neikov, O. D., Naboychenko, S. S., Murashova, I. V., Gopienko, V. G., Frishberg, I. V., Lotsko, D. V. (2009). *Handbook of nonferrous metal powders: technologies and applications* (1st ed.). Elsevier Publisher.

Nouri, A., & Sola, A. (2018). Metal particle shape: A practical perspective. *Metal Powder Report, 73*(5), 276–282.

Yan, J., Zhou, Y., Gu, R., Zhang, X., Quach, W. M., & Yan, M. (2019). A comprehensive study of steel powders (316L, H13, P20 and 18Ni300) for their selective laser melting additive manufacturing. *Metals, 9*(86), 1–20.

Chapter 5
Additive Manufacturing

Additive manufacturing (AM), as defined by the ISO/ASTM 52900 standard, is the process of joining materials to make parts from 3D digital model data. Usually, the material is shaped layer upon layer, as opposed to subtractive (e.g. machining by chip removal) and formative (e.g. plastic deformation) methods of manufacturing. Other historical terms for AM include: 3D printing, additive fabrication, additive layer manufacturing, freeform fabrication, and rapid prototyping.

AM parts are produced using 3D models created by computer-aided design (CAD) software, digital geometries acquired through 3D scanning systems, or even medical imaging equipment. These data are then implemented with instructions for part orientation, for the creation of support structures and slicing strategies, and delivered to the printer for the generation of the real object.

Over the past decade, AM has gained significant attention from designers across a range of industrial sectors including aerospace, energy, automotive, medical, tooling and consumer goods thanks to the ability to create complex geometries with a novel approach, largely free from the constraints of conventional manufacturing routes. To avoid any misunderstanding right from beginning, it must be specified that AM technologies are better suited than others for the production of relatively small lot sizes of complex parts, as schematically shown in Fig. 5.1.

Although the current technology prevents large-scale manufacturing of consumer goods due to cost limitations, the peculiar opportunities offered by AM in terms of ability to create shapes otherwise impossible with conventional processes, the customization of objects without additional costs and the dramatically short lead times (from design to part production) are irreplaceable advantages for specific industrial sectors and markets.

M. Vedani, *Metal Science in Modern Manufacturing Technologies*, Springer Series in Materials Science 351, https://doi.org/10.1007/978-3-031-97125-9_5

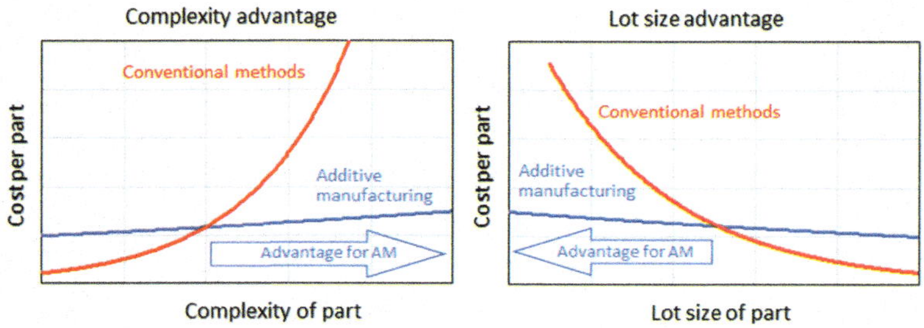

Fig. 5.1 Qualitative evaluation of the advantages offered by AM against conventional manufacturing routes, considering geometrical complexity and lot size of parts. It is shown that AM becomes cost effective for complex parts that are produced in low volumes

5.1 Principles of Metal AM Processes

A large variety of technological processes can be used for AM. The already cited ISO/ASTM 52900 standard lists seven process categories that are sketched in Fig. 5.2 (Vafadar et al., 2021). Powder Bed Fusion (PBF) represents the most popular process owing to its flexibility and ability to produce complex shapes with good geometrical precision. In PBF, a power source (typically a laser or an electron beam) is used to selectively melt a layer of powder material deposited as a thin bed onto a build platform within a sealed chamber. The profile to be melted is defined according to the shape of each "slice" of the 3D object, as defined by the CAD model. After completion of a layer, the build platform is lowered and a new powder layer is deposited and selectively melted onto the previous layers. By the overlapping of a large number of these layers, solid 3D objects are generated (Fig. 5.2a). Once the printing stage is completed, the build chamber is opened, the unmelted powder surrounding the solid part is removed and collected, sieved and stored to be recycled for further use. The printed part is separated from the build platform and subjected to post-processing steps.

In Binder Jetting (BJ) a liquid agent (the binder) is selectively deposited to join the particles on a bed of powder material (Fig. 5.2b). The bound volume is then carefully extracted from the build plate after separation from the surrounding loose powder particles and subjected to several post-print consolidation stages to remove the excess of binder and to achieve the desired welding of metal particles, ruled by the same principles driving the sintering process.

In Directed Energy Deposition (DED), a focused thermal source such as a laser, electron beam or plasma arc is used to melt the feedstock under the form of powder (Fig. 5.2c) or of wire (Fig. 5.2d), while it is deposited using a movable torch driven by a robotic arm. A shielding inert gas is used to protect the molten metal from oxidation and to deliver the powder stream into the molten pool.

Material Extrusion (ME) refers to a processes in which the semi-solid material is selectively dispensed through a nozzle or orifice to shape the part (Fig. 5.2e). For the specific extrusion of metal-based materials, metallic particles are blended with a

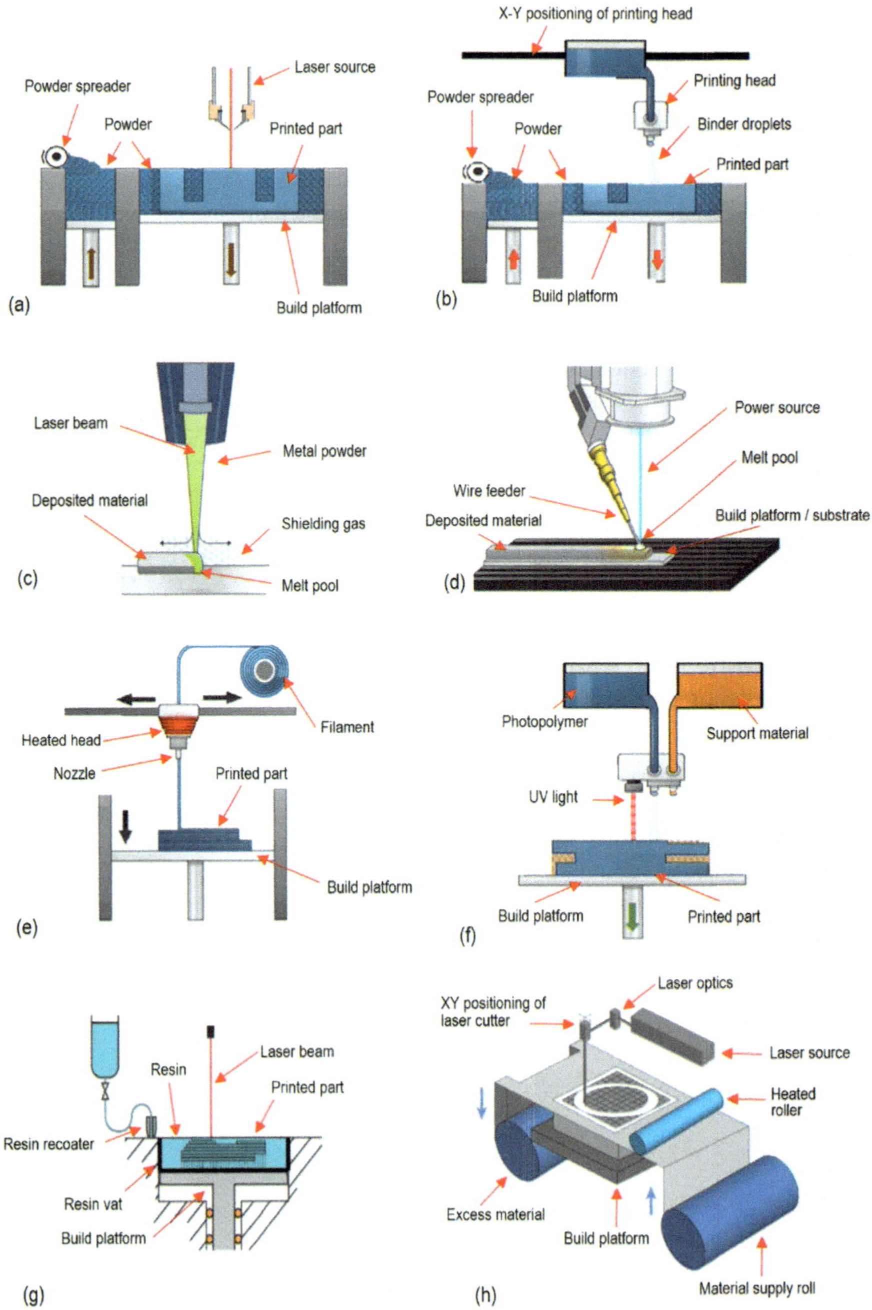

Fig. 5.2 Schematic principles of the main AM process groups (Vafadar et al., 2021). Laser powder bed fusion (**a**), binder jetting (**b**), powder DED (**c**), wire DED (**d**), material extrusion (**e**), material jetting (**f**), vat photopolymerization (**g**), sheet lamination (**h**)

polymeric binder to achieve a viscous behaviour at fairly low processing temperatures and, once the mixture has cured, the part is subjected to post-print sintering, similarly to BJ.

In Material Jetting (MJ) a printhead dispenses liquid droplets of a photosensitive material (thermoset photopolymers) that solidifies under ultraviolet light, building a part layer-by-layer (Fig. 5.2f). Vat Photopolymerization (VP) is a variant of the previous process in which a liquid photopolymer fills a tank (the vat) and it is selectively cured by light-activated polymerization (Fig. 5.2g). Stereolithography is a typical process of this group in which a photosensitive resin is cured by a laser or UV light to build the part layer by layer as the platform lowers into the resin-filled vat.

Finally, in Sheet Lamination (SL), sheets of material are bonded layer by layer to build the desired component (Fig. 5.2h). The sheets are supplied from a feed roller, each sheet is cut to shape with a cutting tool such as a laser to create one layer of the part cross-section. After each layer, the build platform moves down, and a new layer of material is fed. SL is particularly useful when creating parts from prepreg sheets that can easily be bonded in place upon heating, while for metal sheets, ultrasound welding is often adopted to incorporate each new layer onto the 3D part.

5.2 Solidification and Microstructure Development in Metal AM

PBF and DED processes are by far the most common AM processes for metals. They share the same fundamental mechanisms of melting the feedstock material and promoting its solidification under rapid conditions. In both processes, the 3D object is constructed by the accumulation of a large number of partially overlapped tracks with several similarities to a multi-pass welding operation, as shown in Fig. 5.3a.

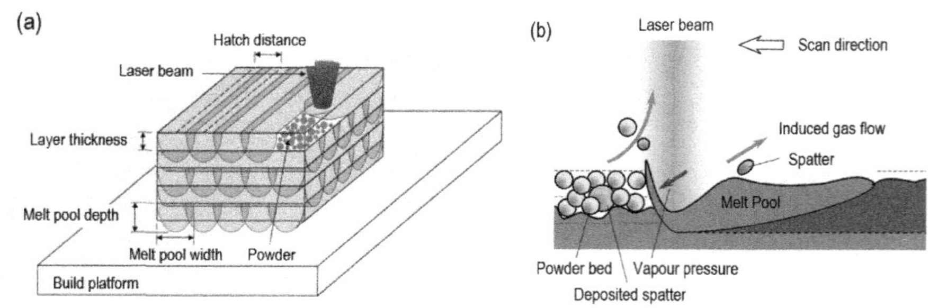

Fig. 5.3 Schematic of the layer-wise deposition of a 3D volume by the scanning action of a laser source (**a**) and detail of the melt pool in LPBF (adapted from Ikeshoji et al., 2022) (**b**)

Indeed, the physics of alloy solidification with these AM processes differs from conventional welding in some peculiar features. The solidification conditions, especially in Laser-PBF (LPBF), are significantly faster owing to the smaller laser spot used and the resulting limited size of the melt pool. The small volume of the melt under solidification from the adjoining track is in contact, and dissipates heat, with already solidified material (the adjacent and underlaying tracks, featuring high thermal conductivity) but also with the powder on the opposite side (featuring reduced conductivity due to gaps among particles and reduced contact area among particles), resulting in an asymmetric thermal field. The interaction of the laser plume and of the metal vapour pressure with the loose powder particles and the melt are responsible of additional phenomena (Fig. 5.3b) like the complex flow of the liquid, the formation of a denudation zone of powder and the ejection of spatters, leading to possible generation of defects (Wang, 2023).

From the material perspective, the melt pool solidification is therefore ruled by the complex interplay of many phenomena including non-equilibrium phase transformations, elemental diffusion (i.e. microsegregation), thermal diffusion, shrinkage effects (i.e. residual stresses), occurring along different length and time scales (Gosh et al., 2023). These phenomena lead to the so-called hierarchical solidification microstructure ranging from the macroscale of the melt pool geometry, to the mesoscale of the grain structure, down to the micro- and atomic scale of the sub-grain microstructure and of the nano-precipitates, as illustrated in Fig. 5.4.

5.2.1 Solidification Conditions

The grain structure and crystal texture of AM parts depend on the solidification conditions experienced within the melt pools, as affected by the heat input provided by the power source and the heat dissipation conditions ruled by the geometrical shape of the melt pool and the thermal properties of the materials involved in the process. The analogy with welding operation can be appropriately exploited here to develop a discussion about the parameters affecting solidification in AM.

In Sect. 3.2.1 (specifically considering Fig. 3.16) a qualitative solidification map was proposed to show the expected morphology and size of the microstructure, based on temperature gradient (G) and growth rate of the solid front (R) found in melt pools. The typical solidification condition corresponding to LPBF can be found next to top-left corner of the diagram, featuring very steep thermal gradients (of the order of 10^6 K/m) and fairly high growth rates of the solid front (of the order of 10^3 mm/s), resulting in high G·R values and promoting the formation of elongated grains containing a fine cellular microstructure. DED processes on the other hand, tend to produce shallower gradients and slower growth rates, resulting in a predominant microstructure formed by coarser columnar-dendritic grains.

During the scanning of the heat source onto the powder bed, both the newly deposited powder layer and a fraction of the already solidified metal belonging to

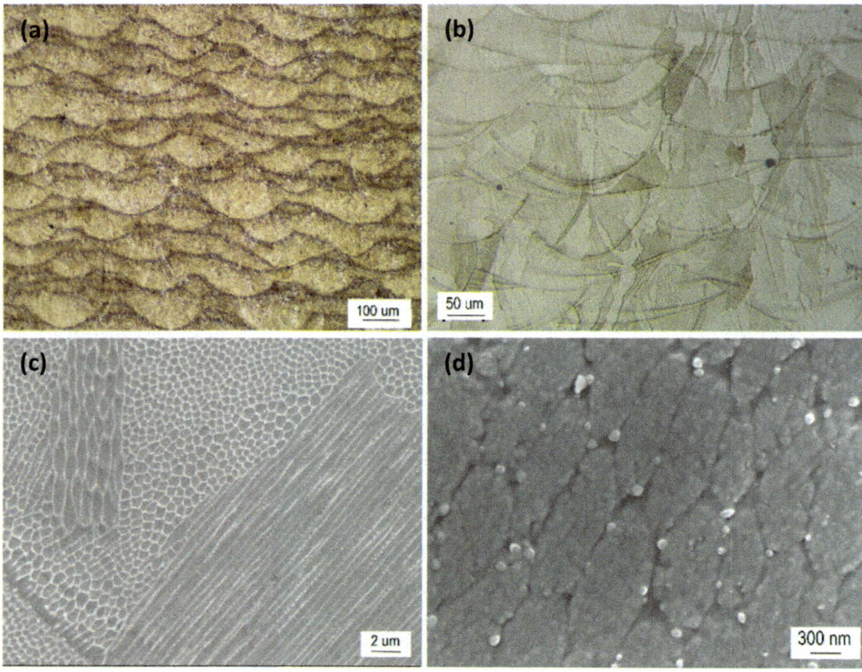

Fig. 5.4 Typical examples of the hierarchical microstructure formed in LPBF metal parts. Molten pool morphology in steels observed from a cross section at different magnifications (**a**) and (**b**); cellular structure formed within the grain volume of a 316L stainless steel (**c**); evidence of precipitation and segregation at cell boundaries in a CM247LC Ni-base alloy (**d**)

previous layers are melted. The exposed solid substrate having the same crystal structure, naturally induces conditions for epitaxial nucleation of the new crystals within the liquid pool, without any need of extensive undercooling or heterogeneous nucleation media. Solidification would then proceed by competitive growth of elongated crystals with their easy-growth direction favourably oriented with respect to heat flow, namely from melt pool boundary toward the centre. Figure 5.5 represents the qualitative shape assumed by the melt pools during LPBF and DED processing, and the expected growth directions of the grains.

Considering that grains will be growing normal to the liquidus isotherm (the blue arrows direction in Fig. 5.5) and by repeated epitaxial nucleation when each new layer will overlap to previously deposited tracks, it can be realized that the growing solid will generally feature columnar grains crossing several melt pools, roughly aligned to the build direction, with a tendency to converge toward melt pool centres and to bend toward the scanning direction when the pool shape becomes deeper and more rounded, as depicted in the example micrograph of Fig. 5.4b. The intrinsic mechanism of grain selection ruled by the easy crystallographic growth direction (the $< 100 >$ direction for bcc and fcc unit cells, and the $< 10\bar{1}0 >$ direction for

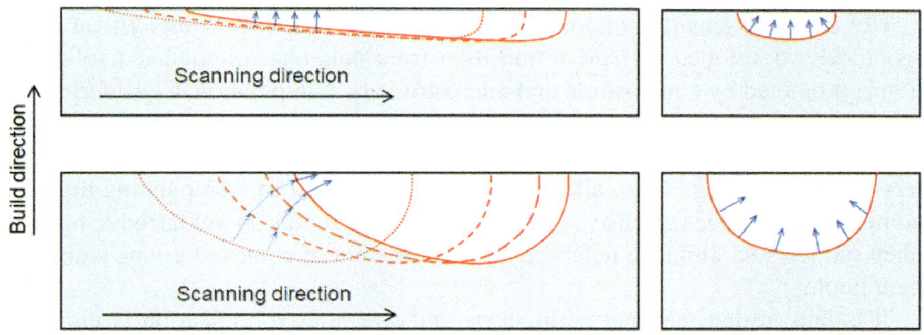

Fig. 5.5 Qualitative representations of the melt pool shape along the longitudinal centerline of the tracks (left) and in the transverse direction (right), for LPBF (top) and DED (bottom) processes. Blue arrows indicate the expected crystal growth direction, which is normal to liquidus isotherm)

hcp unit cells) and by the rather homogeneous heat extraction flow, also lead to the generation of a sharp texture in the columnar grains, resulting in a clear anisotropy of the deformation behaviour of AM alloys (Huangfu et al., 2024).

Recalling the similarity with fusion welding, a further common issue is represented by the sensitivity to hot-cracking of several AM alloys due to the typical textured columnar structure and the severe solidification conditions. As discussed in Sect. 3.2.2, the residual film of liquid located at the grain boundaries can easily promote the development of cracks in the last stages of solidification, especially when coarse columnar grains are formed and they are requested to accommodate shrinkage stresses along their flat and elongated sides, as confirmed by the images in Fig. 5.6.

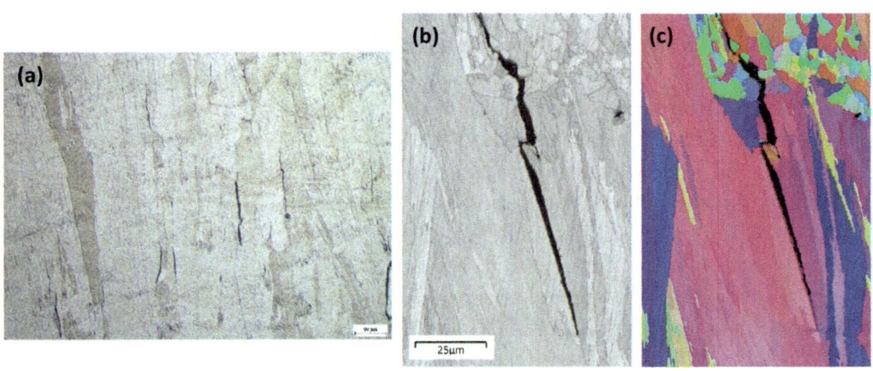

Fig. 5.6 Hot cracking in a LPBF processed CM247LC Ni-base alloy. Optical micrograph of the melt pool structure with evidence of elongated grains and cracks crossing multiple pools (**a**), EBSD images at higher magnification: image quality map (**b**) and orientation map (**c**) of the microstructure next to an intergranular crack

Hot cracking sensitivity is one of the main reasons why in some critical alloys specifically developed for AM a transition from columnar to equiaxed solidification is promoted by composition and microstructure manipulation. Considering that the columnar-to-equiaxed transition (CET) cannot be easily achieved exclusively by a drastic modification of the solidification conditions induced by process parameters (referring to Fig. 3.16), alternative strategies consist in manipulating the alloy composition to induce extensive constitutional undercooling and/or introducing inoculant particles to stimulate heterogeneous nucleation of equiaxed grains within the melt pool.

It is known that the final grain shape and size after solidification is the result of the combination (and competition) between nucleation ability and growth rate of a given alloy under the specific solidification conditions. Undercooling by itself is ruled by several contributions such as thermal, compositional, kinetic effects, and curvature of the solid front. Of these contributions, the compositional effect (constitutional undercooling, discussed in Sect. 1.1.3) plays the biggest role. In this context, a well accepted parameter used to describe the effect of solute elements on the development of the undercooling, is the growth restriction factor Q (Schmid-Fetzer & Kozlov, 2011; Maxwell & Hellawell, 1975). For binary alloys, the strength of distinct elements in promoting constitutional undercooling can be easily evaluated by a simple formula:

$$Q = m \cdot (k - 1) \cdot C_0 \tag{5.1}$$

where m is the slope of the liquidus line on the related binary phase diagram, C_0 is the solute content, and k is the equilibrium partition coefficient. Inserting numerical data in Eq. 5.1, one can collect information about the ability of specific solute elements to induce constitutional undercooling during fast melting, thus promoting CET.

Besides undercooling, inoculants are frequently used both in casting alloys and in AM alloys to stimulate the formation of equiaxed grains. The appropriate dispersion in the melt of tiny solid particles (externally added or in-situ formed) with lattice parameters as close as possible to those of the solidifying alloy can effectively lower the energy barrier for the heterogeneous nucleation of grains, partially or totally suppressing the epitaxial nucleation and growth of the columnar structures. The efficiency of a nucleant agent depends on several factors like the interfacial energy between the nucleant and the melt, which rules the wetting behaviour, its crystal structures and chemical composition (Tan & Zhang, 2024). The control of solidification by inoculants is widely exploited in Al alloys. Inherited from foundry alloys, is the combined use of Al, Ti and B to produce in situ at high temperature TiB_2 and Al_3Ti dispersoids, that can be effective as seeds for new equiaxed grains. Other Al-based feedstock powders for AM rely on the addition of small quantities of elements like Zr, Ti, Sc, Ta, Er to form Al_3X particles (being X one of these elements) with L1-2 ordered crystal structure.

Figure 5.7a graphically shows the evolution of phases during solidification of an Al–Cu–Mg alloy inoculated with TiB_2 particles during LPBF. At elevated temperatures, solid TiB_2 particles already exist within the liquid melt. When approaching the liquidus temperature of the Al alloy, further Al_3Ti can form, both stimulating the formation of small equiaxed grains. Figure 5.7b depicts TEM images of the grain structure with evidence of the described particles at grain cores.

Figure 5.8 is a further example describing the progressive transition from columnar to equiaxed structure investigated by using a reference Al-5Mg alloy whose composition was modified with increasing fractions of inoculants.

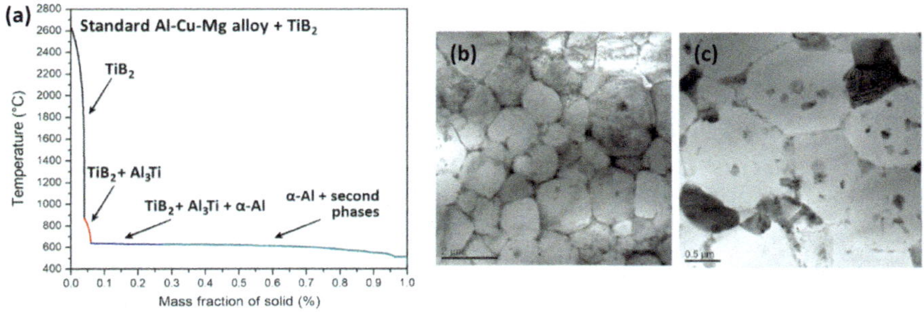

Fig. 5.7 Calculated phase evolution during solidification of an Al–Cu–Mg alloy containing TiB_2 inoculants (**a**) and ultrafine equiaxed grain structure with embedded type TiB_2 nucleating particles (**b** and **c**) (courtesy of A. Fabrizi, Università di Padova)

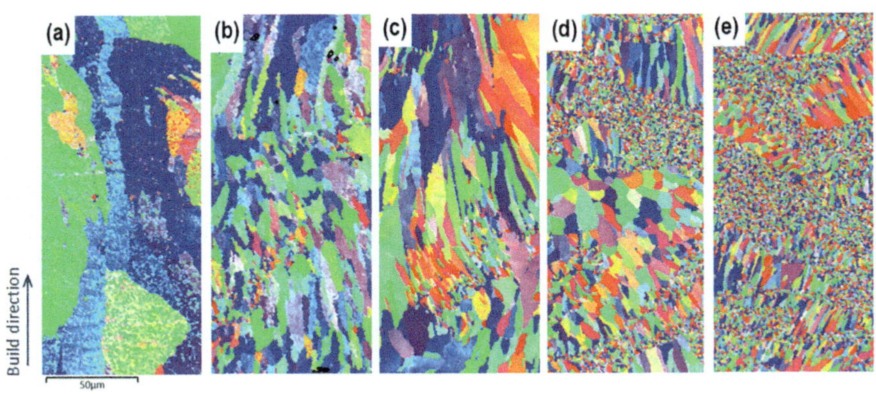

Fig. 5.8 EBSD orientation maps of LPBF processed Al-5Mg alloys containing different amounts of inoculants. (**a**) undoped Al-5Mg alloy; (**b**) Al-5Mg alloy with reduced amounts of Sc and Zr; (**c**) Al-5Mg alloy with reduced amounts of Sc and Ti; (**d**) commercial Al-5Mg alloy doped with Sc and Zr; (**e**) same as (**d**) with increased amount of Zr (adapted from Larini et al., 2024)

5.3 Post-processing of AM Parts

Post-processing represents a significant step in the manufacturing cycle of AM parts and can add to a large fraction of the total cost of the finished components. Typical operations include powder removal (for powder-bed processes), stress relieving, separation of the part from the building platform and cutting the supports, surface finishing, machining of selected functional surfaces, hot isostatic pressing and heat treatments to improve properties.

5.3.1 Development of Residual Stresses and Stress Relaxation Treatments

Residual stresses can be defined as the stresses that remain in a material under equilibrium, when any external load is removed. The presence of residual stresses must be considered for the safe manufacturing and accurate prediction of the service life of any AM component. Indeed, part distortion, loss of geometric tolerances, delamination, cracking, premature fatigue failure can all be deeply affected by the residual stress field generated during AM processing (Xie et al., 2022, Bartlett, 2019). A representative example of part delamination due to residual stresses in a massive part produced by LPBF is depicted in Fig. 5.9.

The origin of residual stresses in AM can be intuitively understood considering a simplified model representing the layer-wise building process which develops by the repeated deposition of liquid layers on top of the already consolidated material. Figure 5.10.a schematically depicts this simplified mechanism, often referred to as the temperature gradient model (TGM) in the literature (Xie et al., 2022). Right after solidification of a new layer, the top portion of the build lies at a higher temperature than the lowermost volume. The former layer will try to expand (being restricted due to continuity with the cooler underlying part), while the latter will be stretched by

Fig. 5.9 Delamination along deposited layers in a bulky LPBF Maraging steel part

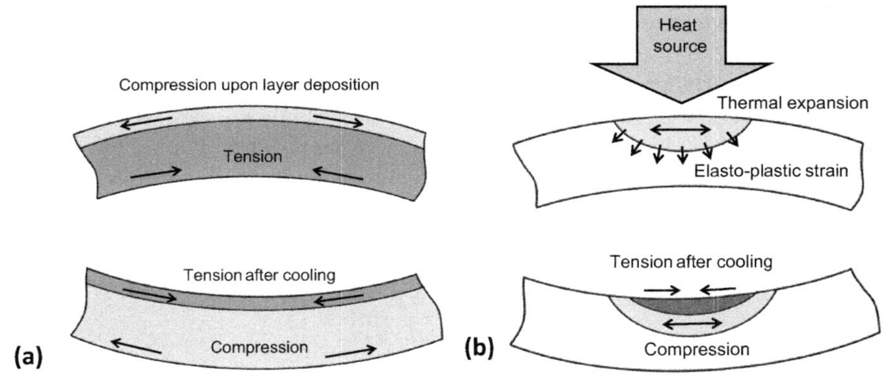

Fig. 5.10 Simplified mechanisms for the development of residual stresses and distortions in layer-wise AM processes. (**a**) hypothesis of entire layer instantaneously melted; (**b**) hypothesis of confined melt pool

the more expanded top layer. As a result, compressive stresses will develop in the new layer and tensile stresses in the underlying volume. Once the heat source will leave the spot, the last layer will cool quickly from a higher temperature, contracting at a greater rate than the volume beneath it, eventually generating a state of tensile stress in the top layer and compressive stress in the underlying volume.

For sake of precision, the basic mechanism described above should be refined to better reflect reality, taking into account that the new layers are not deposited instantaneously and homogeneously. Instead, a confined melt pool is formed and moves along a specific path defined by the scan strategy of the heat source. This leads to more complex thermal gradients resulting in elastic and elasto-plastic strain gradients in space and time (see Fig. 5.10b). Moreover, the repeated deposition of new layers induces on the consolidated material a series of thermal cycles featuring a decreasing amplitude with increasing distance from surface, which adds complexity to the final stress field, due to a combination of newly generated thermal gradients and of annealing effects on the material.

The distribution and amplitude of residual stresses depend on a wide range of factors. Thermal and mechanical properties of the materials are of paramount importance. As a general rule, it can be stated that materials with higher diffusivity can better mitigate the effects of thermal strains (during quenching, welding and AM). The thermal diffusivity D_t is linked to other physical properties by Eq. 5.2:

$$D_t = \frac{k}{\rho \cdot c} \tag{5.2}$$

where k is the thermal conductivity, ρ is the density and c is the specific heat. Therefore, materials featuring higher thermal conductivity, lower density and lower specific heat, would better accommodate the thermal effects. In addition, the coefficient of thermal expansion, α, which translates the thermal gradients into strain gradients plays and important role: the lower α, the limited the distortion and residual stress amplitude.

Besides material properties, the mechanism of residual stress build-up driven by the intense thermal gradients and cooling rates, is strongly affected by the selected AM technology and by its process parameters. Energy input is the most obvious parameter to be evaluated when assessing the overall effect of several process variables. In powder bed AM processes, the heat input is typically evaluated by the volume energy density (VED) E_V defined as:

$$E_V = \frac{P_B}{h \cdot d \cdot v_S} \tag{5.3}$$

where P_B is the beam power, h is the hatch distance separating adjacent tracks, d is the layer thickness, and v_S is the scan speed. Although the equation only gives a rough estimation of the actual energy density supplied to the processed material, there is consensus on suggesting that the lowest possible VED that can produce fully dense parts should be adopted to minimize residual stresses and distortions.

Among possible strategies to mitigate residual stresses, feedstock preheating is one of the most efficient methods. As with welding processes, it can be recognized that increasing the temperature of the powder bed substrate can lower the temperature gap between top and bottom regions, hence the thermal strains. Preheating of the powder can be achieved by implementing a preheated build platform in LPBF systems, while in electron-beam based processes, in-situ heating of the powder can also be easily achieved by a rapid low-energy scanning of the whole bed before performing the selective melting scan.

Post process heat treatments are an additional strategy to control residual stresses. Indeed, once the build platform has been extracted from the AM system, a first common treatment to be done before freeing the printed parts from its supports involves residual stress relaxation by annealing. Depending on alloy composition and target properties, suitable combination of annealing times and temperature should be selected. Additionally, it is to mention that post treatments can concurrently be designed for other purposes as well, such as tempering the martensite formed due to fast cooling in alloy steels, or promoting the precipitation of supersaturated solute elements to form nano-size strengthening phases in many non-ferrous alloys.

5.3.2 Surface Features and Finishing Treatments

The poor surface quality of AM parts is one of the main limitations of this technology, implying additional costs and efforts for its improvement in case of structurally and aesthetically demanding applications.

Figure 5.11 shows representative images of the surface morphology of as-processed LPBF parts. Views from the upper side representing the last layer that was melted before stopping the process (Fig. 5.11a) gives clear evidence of the parallel

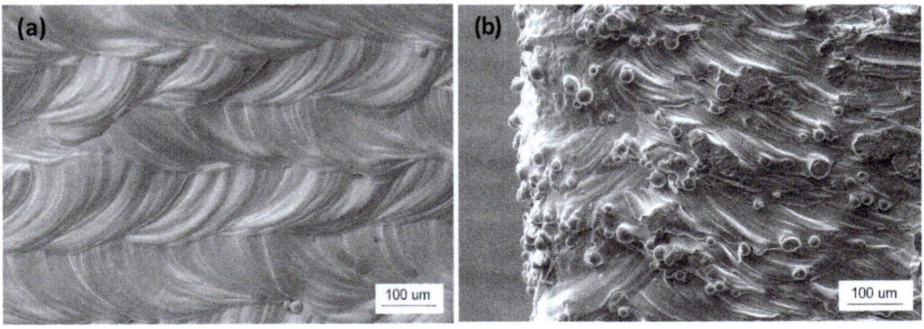

Fig. 5.11 Surface morphology of pure copper samples printed by LPBF; (**a**) top surface, (**b**) lateral view

adjacent tracks, with partial overlapping of the melt pools along alternate directions. The lateral view in Fig. 5.11b additionally highlights the presence of a large number of partially melted powder particles stuck on the surface. These features accounts for the poor surface roughness reached in AM parts, which is typically in the range Ra = 5–15 μm for conventional LPBF processes.

There exist several physical phenomena that can contribute to generate surface asperities on AM parts (DebRoy, 2018; Snyder, 2020; Viale, 2022). To begin with, a first effect is caused by the layer-wise fabrication process typical of AM. A so-called "stair step" effect is produced by the approximation of inclined and curved surfaces with stepped shifted layers. Reduction of the layer thickness can mitigate this effect, improving the surface quality, but only at expenses of longer processing times. Other effects, roughly on a similar length scale, involve the shape of the melt pools and the hatch spacing between adjacent tracks. Unproper setting of these process parameters can form large ridges between passes (see Fig. 5.11a) depleting the surface quality. This effect is mostly pronounced on the upskin surfaces (upward facing surfaces) of parts, but can also affect the lateral surfaces. On the other hand, downskin surfaces are deeply affected by the reduced thermal conductivity of the loose powder particles forming the bed. When the accumulated heat produced along the first unsupported layers does not dissipate efficiently and homogeneously, the bottom of the melt pools resting on loose powder particles generates uneven surfaces and dross protuberances irregularly penetrating the powder bed, leading to poor surface quality.

A second group of mechanisms affecting AM surfaces involves incomplete melting of powder particles and balling phenomena. At the edges of the printed parts, the laser track creates a roughly defined interface where some exposed powder particles can only be partially melted or they get sintered to the consolidated volume (see Fig. 5.11b). In addition, the engulfment of spatters plays an important role. Recalling, Fig. 5.3b, spatters are produced by the ejection from the melt pools of liquid droplets that can solidify during their flight and land on powder bed surface. When their size exceeds that of the powder, they can hardly be engulfed and remelted in the subsequent layers, inducing the risk of damage for the recoater and promoting the generation of lack of fusion defects. Besides, those spatters lying close to the

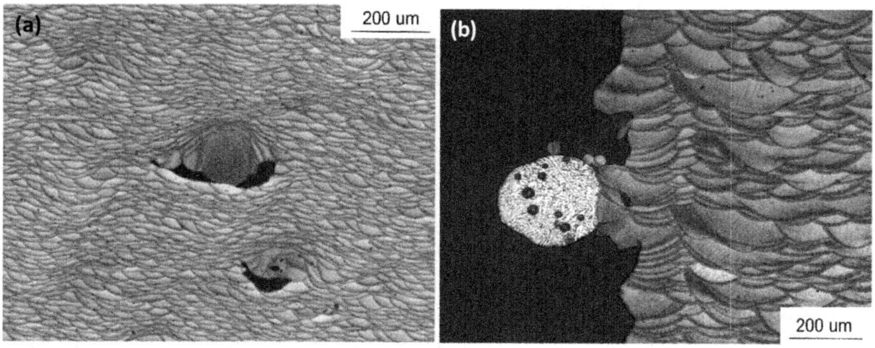

Fig. 5.12 Large-size spatters embedded in the as built microstructure (**a**), and located at lateral surface (**b**) of an Al-7Mg–Si alloy sample

edges of the sample will be only partially remelted during the contour scans and will remain attached alongside the surface of the as-built part, impairing surface roughness.

Figure 5.12 depicts some of the above-described phenomena. Figure 5.12a represents coarse spatters which have been embedded by the following layers in the microstructure formed by LPBF. Their large size was such that they could only partially be remelted and the powder from the following layers could not be properly spread close to their position, thus generating cavities (lack of fusion defects). Figure 5.12b shows the effect produced by a spatter next to the lateral surface of a part.

The images also reveal that spatters can be clearly distinguished from the normal microstructure by their coarser dendritic structure (in this case made of primary α-Al and inter-dendritic α-Al/Si eutectic), resulting from the slower cooling rate experienced during their flight in the build chamber while airborne, driven only by convection and radiation (Nasab et al., 2018).

Finally, balling phenomena can appear on part surfaces. It has been observed that at high scanning speeds the melt pools become elongated and eventually can break into small portions due to Plateau-Rayleigh instability. Due to surface tension effects, these small droplets will assume a spherical shape and segregate at the edges of the tracks, creating protrusions when exposed to surfaces of the printed part.

Given the variety of anomalies and the rough profile of AM surfaces, post processing is more the rule than the exception in high-performance parts. Surface finishing requirements are dictated by the final application. For example, in biomedical implants, heat exchangers, and in general for fluid handling, it is important to avoid occasional loss of powder particles poorly adhering on surfaces that could contaminate the fluids. Sand blasting, often followed by electropolishing, are typically considered for these situations. For structural parts subjected to fatigue loading, recesses on surface profile should be especially controlled. In this case, post-processing techniques such as mechanical rumbling (relative movement of the parts within a vibrating barrel filled with grinding media), often assisted by reactive fluids can conveniently be used. Conventional shot peening, laser shock peening, and laser polishing have also been evaluated to improve surface quality, even though the limitations dictated by the accessibility of the media in recesses, cavities and internal channels should always be considered (Mahmood et al., 2022).

References

Bartlett, J.L., Li, X. (2019) An overview of residual stresses in metal powder bed fusion. *Additive Manufacturing, 27*, 131–149.

BS EN ISO/ASTM 52900. (2021). Additive manufacturing—General principles—Fundamentals and vocabulary standard.

DebRoy, T., Wei, H. L., Zuback, J. S., Mukherjee, T., Elmer, J. W., Milewski, J. O., Beese, A. M., Wilson-Heid, A., De, A., & Zhang, W. (2018). Additive manufacturing of metallic components—Process, structure and properties. *Progress in Materials Science, 92*, 112–224.

Ghosh, S., Zollinger, J., Zaloznik, M., Banerjee, D., Newman, C. K., & Arroyave, R. (2023). Modeling of hierarchical solidification microstructures in metal additive manufacturing: Challenges and opportunities. *Additive Manufacturing, 78*, Article 103845.

Huangfu, B., Liu, Y., Liu, X., Wu, X., & Bai, H. (2024). Anisotropy of additively manufactured metallic materials. *Materials, 17*, 3653.

Ikeshoji, T. T., Yonehara, M., Kato, C., Yanaga, Y., Takeshita, K., & Kyogoku, H. (2022). Spattering mechanism of laser powder bed fusion additive manufacturing on heterogeneous surfaces. *Scientifc Reports, 12*, 20384.

Larini, F., Casati, R., Marola, S., Andrianopoli, C., Bajaj, P., & Vedani, M. (2024). Effect of combined additions of Sc, Zr and Ti on hot-cracking resistance and precipitation behaviour in Al-Mg alloy by L-PBF. *Materialia, 35*, Article 102127.

Mahmood, M. A., Chioibasu, D., Rehman, A. U., & Popescu, A. C. (2022). Post-processing techniques to enhance the quality of metallic parts produced by additive manufacturing. *Metals, 12*, 77.

Maxwell, I., & Hellawell, A. (1975). A simple model for grain refinement during solidification. *Acta Metallurgica, 23*(2), 229–237.

Nasab, M. H., Gastaldi, D., Lecis, N. F., & Vedani, M. (2018). On morphological surface features of the parts printed by selective laser melting (SLM). *Additive Manufacturing, 24*, 373–377.

Schmid-Fetzer, R., & Kozlov, A. (2011). Thermodynamic aspects of grain growth restriction in multicomponent alloy solidification. *Acta Materialia, 59*, 6133–6144.

Snyder, J. C., & Thole, K. A. (2020). Understanding laser powder bed fusion surface roughness. *Journal of Manufacturing Science and Engineering, 142*, Article 071003.

Tan, Q., & Zhang, M. (2024). Recent advances in inoculation treatment for powder-based additive manufacturing of aluminium alloys. *Materials Science and Engineering, R, 158*, Article 100773.

Vafadar, A., Guzzomi, F., Rassau, A., & Haywar, K. (2021). Advances in metal additive manufacturing: A review of common processes, industrial applications, and current challenges. *Applied Sciences, 11*, 1213.

Viale, V., Stavridis, J., Salmi, A., Bondioli, F., & Saboori, A. (2022). Optimisation of downskin parameters to produce metallic parts via laser powder bed fusion process: An overview. *The International Journal of Advanced Manufacturing Technology, 123*, 2159–2182.

Wang, J., Zhu, R., Liu, Y., Zhang, L. (2023) Understanding melt pool characteristics in laser powder bed fusion: An overview of single-and multi-track melt pools for process optimization. *Advanced Powder Materials*, 2 100137.

Xie, D., Lv, F., Yang, Y., Shen, L., Tian, Z., Shuai, C., Chen, B., & Zhao, J. (2022). A review on distortion and residual stress in additive manufacturing. *Chinese Journal of Mechanical Engineering: Additive Manufacturing Frontiers, 1*, Article 100039.

Chapter 6
Metallurgy of Subtractive Processes

Subtractive manufacturing encompasses a wide group of technologies where objects are created starting from a solid block by progressively removing material. It includes a wide group of technologies such as turning, drilling, milling, as the main representatives of traditional chip-removal machining processes, together with laser cutting, waterjet cutting, plasma cutting, electro-discharge machining, chemical milling. Quite often, the shape to be produced is defined by a CAD model and the tools are computer numerically controlled (CNC machining) to achieve the desired geometry with high accuracy. The flow chart in Fig. 6.1 proposes a general classification of the subtractive processes based on material removal (Davis, 1998).

Subtractive manufacturing technologies apply to a broad range of materials (metals, polymers, composites, ceramics and many natural materials like wood or rocks) with high flexibility in terms of achievable shapes, dimension and surface finish. As opposed to additive manufacturing (see Chap. 5) where the material is added to the designed shape only where it is needed, subtractive processes generate chips or smaller particles starting from a blank of larger volume than the part to be obtained. This excess of material is considered as a waste even though chips can be recycled in most of the cases. Despite this apparent inconsistency toward the search for sustainable processes, subtractive processes are still, and by far, the reference manufacturing route in many industrial fields owing to cost effectiveness, overall energy savings and short lead times, especially when suitable semi-products are readily available on the market.

In this section the effect of material properties on machinability will be discussed. The main focus will be given on conventional chip-removal machining processes for metallic alloys and on how their microstructure and properties can affect the process performance, considering the fundamental mechanism of material removal and the ability to mitigate the wear of the machining tools. Further considerations will be extended, considering material-related aspects, to abrasive and few other non-traditional processes.

M. Vedani, *Metal Science in Modern Manufacturing Technologies*, Springer Series in Materials Science 351, https://doi.org/10.1007/978-3-031-97125-9_6

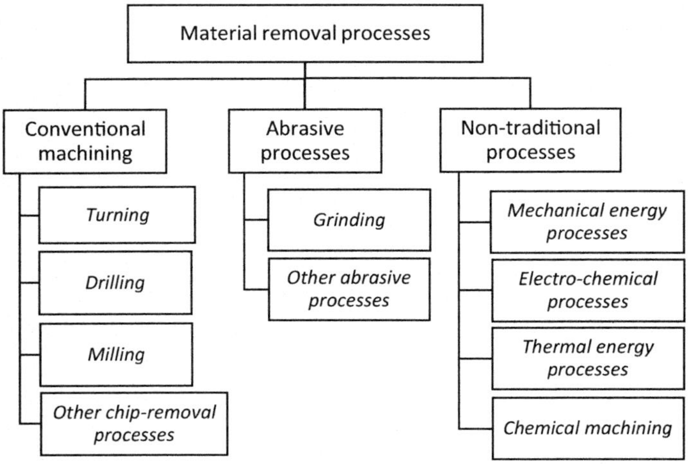

Fig. 6.1 General classification of the subtractive processes

6.1 Principles of Conventional Machining

6.1.1 Chip-Based Machining Processes

Most of the conventional machining processes are based on the concept of removing the surplus of material from a starting block by using tools in relative motion to the surface of the workpiece, as schematically depicted in Fig. 6.2. The excess of material is gradually removed as a chip, which flows ahead of the tool edge and leaves the working area.

The basic principle of chip formation can be simplified as the cutting action performed by the moving tool which travels slightly below the surface level of the workpiece, with a defined depth of cut, as represented in Fig. 6.3. The chip is separated from the workpiece once it approaches the tool edge and slides with friction

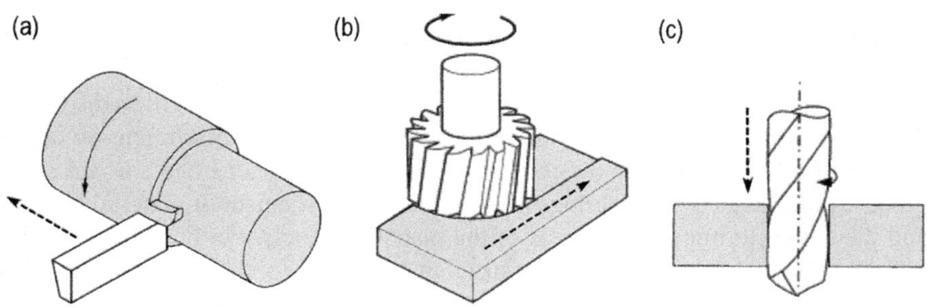

Fig. 6.2 Schematics of some conventional machining processes; **(a)** turning, **(b)** milling, **(c)** drilling

along a certain distance onto the rake surface, before leaving the tool. This oversimplified configuration requires a number of further clarifications when moving to real machining processes. To start with, along the third direction we can consider that the cutting-edge line of the tool can quite often be oblique to the cutting direction (especially in drilling and milling), thus introducing a more complex 3D flow of the chip in the working area. The rake angle may vary over a wide range, including negative values, and the rake surface often differs from a planar surface to better control the flow of the chip and its separation from the tool.

Abrasive processes, like grinding, honing, lapping, differ from the above technology in that small material particles are removed from the workpiece by an abrasive mechanism promoted by a tool that contains hard particles, the abrasive grains. The allowable material removal rates are generally lower, but closer tolerances and more accurate surface finish can be achieved. The simplified material removal mechanism for abrasive processes is depicted in Fig. 6.4. The cutting effect is promoted by a large number of abrasive grains that simultaneously interact with a thin layer of the workpiece material. The tiny volumes of materials that are caught among adjacent abrasive particles produce fragments that are expelled from the working area under the effect of the relative movement of the grinding wheel and of a flushing agent.

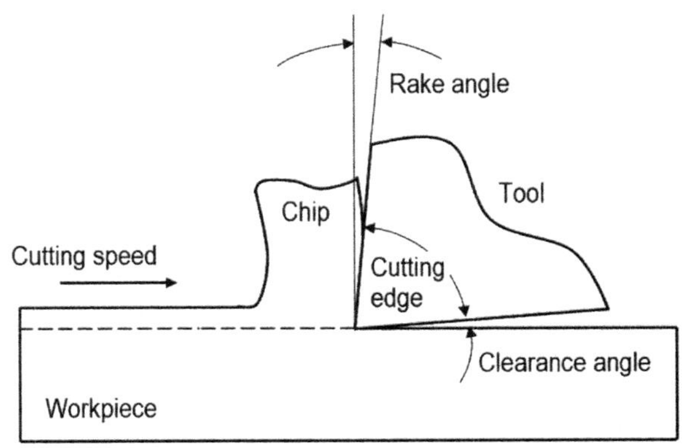

Fig. 6.3 Simplified mechanism of material cutting to produce the chip during machining

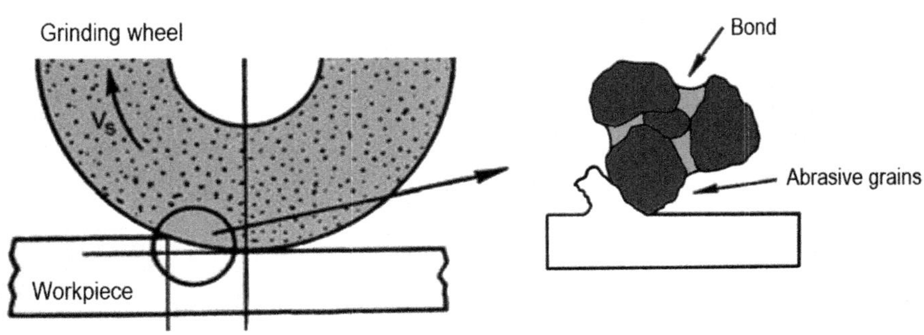

Fig. 6.4 Simplified mechanism of material removal by abrasive grinding

6.1.2 Mechanism of Chip Formation

Considering now the principle of chip formation from the material perspective, it is assumed that during conventional machining the chip is formed by virtue of extensive shear deformation, as depicted in Fig. 6.5a. When the material located within the layer of the depth of cut approaches the tip of the tool, it is plastically deformed by shear to abruptly change its flow direction, driven by the rake angle of the tool. The shearing of the material takes place within distinct and relatively narrow bands (inclined by a specific shear angle that depends on cutting tool geometry, process parameters and material behaviour). While sliding along the rake face of the tool, the so-formed chip is subjected to additional shear deformation (the secondary shearing) close to contact surface, due to frictional effects. Heat due to friction and extensive deformation of the chip at a very fast rate are also produced, which results in an increase of temperature and accelerated wear of the tool.

Depending on material properties and on machining parameters, the chip produced may assume different shapes. In ductile metals and alloys, large plastic deformations can readily be accommodated and a continuous chip is generally produced (see Fig. 6.5b). As a result, the chip is in contact with the rake face of the tool for a longer distance, thus increasing the amount of friction, the mechanical work associated

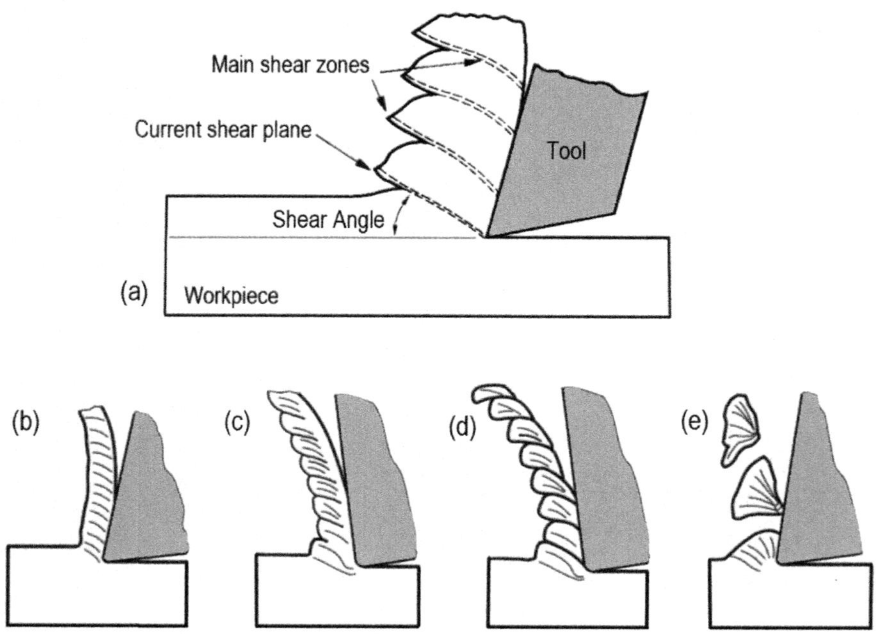

Fig. 6.5 Chip generation by shear mechanism; **(a)** the position of shear plane and of the main shear zones, **(b)** aspect of continuous chips, **(c)** lamellar chip, **(d)** segmented chip, and **(e)** discontinuous chip

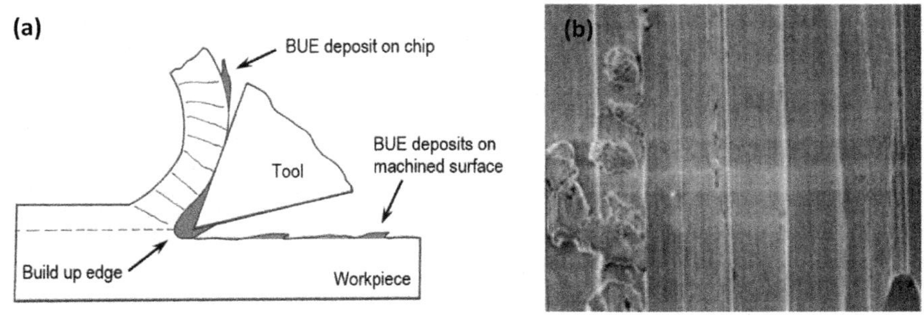

Fig. 6.6 Mechanism of formation of the BUE (**a**) and view of BUE deposits on a machined surface (**b**)

to machining, and the heat to be dissipated. Other alloys with reduced ductility or even with a brittle behaviour tend to produce chips that progressively shift to a lamellar aspect, a segmented shape, and eventually become discontinuous, consisting of fine fragments as a result of the periodic brittle failure of the processed material (see Figs. 6.5c–e). The latter mechanism is the most appreciated for conventional machining operations since it reduces tool wear and provides accurate finishing and size control.

In ductile alloys, when the pressure on the tool face exceeds certain limits, the increased local temperature may even promote the accumulation and temporary welding of small portions of workpiece material at the rake surface and at the cutting edge, thus creating a built up edge (BUE). Not only the presence of a BUE modifies the geometry of the tool, but its intrinsic instability produces periodic separation of these portions which can stick to the chip or slide beneath the tool edge to adhere to the machined surface, as shown in Fig. 6.6. The presence of the BUE therefore leads to reduced tool life, poor surface finish and uncontrolled dimensional tolerances.

6.2 Machinability of Alloys

6.2.1 Effects of Microstructure

Machinability can be defined as the ease with which materials can be machined under a given set of process conditions including type and geometry of tool, cutting speed, feed rate and depth of cut. Quantifying machinability can be ambiguous since many parameters can be used to measure it. Tool wear rate or tool life are of greatest interest for practical applications whereas other criteria often adopted for scientific investigations consider the machining forces and power consumption, the morphology of the chips, the achieved surface roughness. Irrespective of the evaluation criterium, there exists full agreement about the strong influence of material properties, especially microstructure, on machinability (Ezugwu, 2003; Schultheiss, 2017; Stavroulakis, 2022; Zhao, 2022).

Poor machinability is generally associated to one or more of the following material characteristics:

- high strength and hardness, especially at high temperature (during chip formation), which increases wear rate and the possibility of deformation or failure of the tool;
- high strain-rate sensitivity, resulting in higher cutting forces and shear localization;
- elevated work-hardening rate of the alloy, which results in a harder material after shearing and a rapid wear of the tool face and cutting edge;
- low thermal diffusivity, leading to overheating at tool tip and inducing sharper thermal gradients;
- tendency to weld or seize to rake face and cutting edge of the tool, thus promoting the formation of BUE.

When specifically considering the alloy microstructure, there exist several general concepts that can help evaluating the machinability of alloys. Hard phases dispersed in the matrix are for obvious reasons deleterious for tool life. Carbides and oxide inclusions in steels, other intermetallic particles in several non-ferrous alloys, ceramic reinforcement in metal matrix composites, they all deplete machinability by hindering the deformation of the chip and reducing tool lives due to flank wear. To some extent, their effect on wear of the tool can be compared to the action of the abrasive grains in the grinding wheel depicted in Fig. 6.4. When exposed to the chip surface, hard particles are pushed and forced to slide against the tool face and produce an abrasive wear effect. Machinability is therefore affected by the volume fraction of the hard particles, by their size and distribution, and by their hardness. More specifically, it has been stated (Schultheiss, 2017) that the abrasiveness of a particle-containing alloy can be evaluated using Eq. 6.1:

$$W_{ab} = H + V_{fII} \cdot (H_{II} - H_I) \tag{6.1}$$

being W_{ab} an abrasiveness parameter of the alloy, H the average hardness of the alloy, V_{fII} the volume fraction of the second phase hard particles, H_{II} and H_I the microhardness of particles and matrix, respectively. This equation demonstrates that the abrasiveness of a complex alloy also depends on the hardness difference between the different constituents.

In simpler alloys that can be approximated as single-phase, it is reasonable to assume that strength and plasticity of the main constituent play the major role in machinability. From one side, low-strength materials result in lower cutting forces and reduced abrasiveness for the tools, However, when the low strength is associated to extensive ductility of the alloy, concerns may arise due to generation of continuous chip and BUE. Grain size is a further important microstructural parameter affecting machinability. It is well known that a reduction in grain size results in higher strength and hardness, that generally shorten tool lives and increase cutting forces. However, considering the fundamental mechanisms of chip generation based on shear plastic strain ahead of the cutting edge (see Sect. 6.1.2), it can be speculated that the ease and repeatability by which this mechanism can take place will depend on the specific crystal orientation of the group of grains (or the single grain) exposed to the tip of the tool and to the ideal shear angle (see Fig. 6.5a). When machining coarse-grain microstructures such those found in castings, or in case of high-precision machining where the depth of cut can be as low as 0.05 mm, the grain size can become comparable to the depth of cut and thickness of chip. As a result, the periodical change of orientation of the few grains exposed during the cutting process can cause fluctuation in cutting forces, poor surface finish, and reduced accuracy in dimensional tolerances.

6.2.2 *Free Machining Alloys*

Given the effects of microstructure on machinability described above, it is expected that alloys can be specifically designed to meet those requirements that will ease machinability, such as discontinuous chip formation with lower cutting forces and reduced friction on the tool. Such materials are conventionally named as free-machining alloys. It can be observed that in some cases, free-machining alloys are found as a suitable combination of constituents in already existing materials, typical examples are graphite-bearing cast irons. In other cases, variants of already existing structural alloys have been deliberately designed to achieve improved machinability, such as in lead-bearing brasses or in resulphurized steels.

Graphite is the main second phase of many cast irons. It can assume the shape of flakes, of compacted (also called vermicular) graphite or of spheroidal graphite that is embedded in a Fe-based matrix consisting of ferrite, pearlite or less frequently bainite and martensite. Figure 6.7 shows representative microstructures of flake and spheroidal cast irons.

The atomic structure of graphite consists of hexagonal rings of carbon atoms covalently bonded in a planar configuration, that are stacked parallel to each other

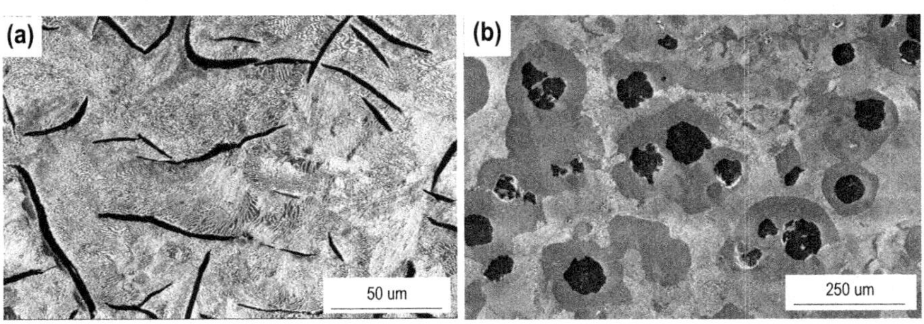

Fig. 6.7 Typical microstructure of flake grey cast iron with pearlitic matrix (**a**) and of spheroidal cast iron with ferritic-pearlitic matrix (**b**)

by weaker Van Der Waals bonds. This arrangement is responsible for the excellent thermal and electrical conductivity, for the brittleness and poor strength of graphite. Besides, the easy shearing of the stacked planes also brings peculiar self-lubricating properties under sliding conditions, that represent a remarkable advantage in machining. Indeed, during chip formation of graphitic cast irons, the graphite particles exposed at the tool edge and along the rake surface will produce a so-called tribofilm between the two mating surfaces, thus reducing friction and cutting forces. Finally, it is to observe that the microstructural discontinuities induced by graphite promote the formation of finely fragmented chips. It can be stated that machinability generally ranks with the following shape of graphite (from worst to best): spheroidal, compacted, flake.

Lead, as a deliberately added element in several ferrous and non-ferrous alloys, can act in a similar way as graphite in cast irons. Although health hazards and progressively more restrictive regulations are encouraging alternative strategies for free-machining materials, lead-bearing alloys are still in use in many sectors where efficiency in machining operation is of major importance.

Lead is substantially insoluble at the solid stage in most structural alloys such as steels, aluminium and copper alloys. Therefore, the lead-bearing free-machining alloys show a microstructure consisting of dispersed lead particles located mainly at grain boundaries as a result of the segregation of these low-melting point droplets in the last stages of solidification. Leaded brass could be taken as a representative example of these materials. The phase evolution during solidification of a Cu37Zn3Pb alloy and the typical microstructure of a wrought bar are shown in Fig. 6.8. From the phases predicted by a simulation using the Calphad method based on equilibrium transformations (see Fig. 6.8a), it can be inferred that from the high-temperature liquid where Pb is fully soluble, a primary β-phase forms that partially decomposes at lower temperature into an α-phase to produce the expected biphasic brass microstructure. Pb is insoluble in these solid phases and segregates right after the solidification of the brass matrix in a separate liquid that will later solidify at the temperature of 327 °C. The obtained microstructure is depicted in Fig. 6.8b to highlight that the preferred sites of Pb accumulation are the α/β interphase boundaries

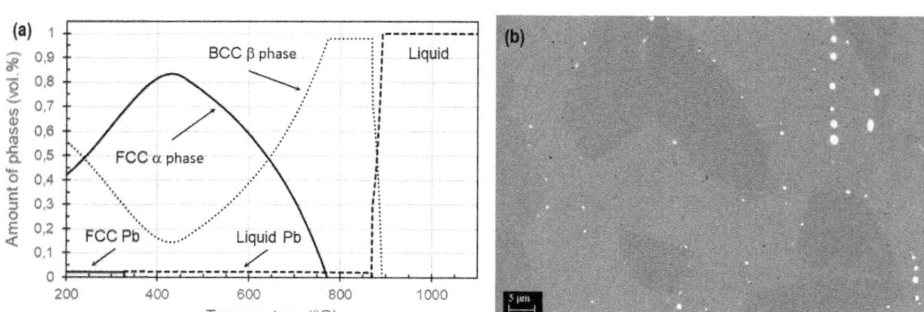

Fig. 6.8 Phase evolution during solidification and cooling of a Cu39Zn3Pb alloy calculated under equilibrium (**a**) and resulting microstructure (**b**)

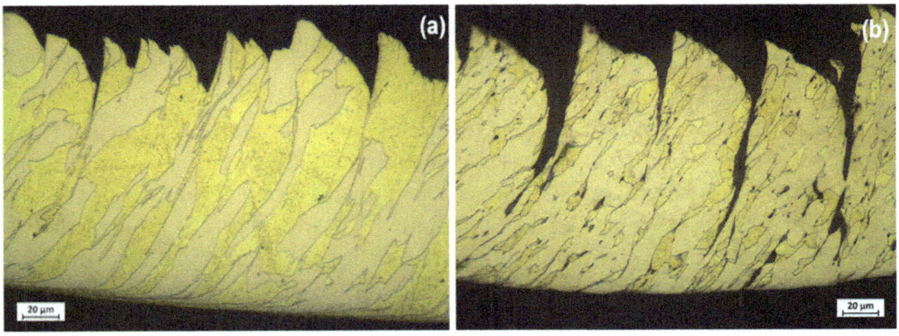

Fig. 6.9 Microstructure and shape of chips produced by turning a standard lead-free Cu37Zn alloy (**a**) and a free-machining Cu37Zn3Pb brass (**b**)

and that Pb, when at the liquid stage, tends to reduce its surface energy assuming a spherical shape, resulting in a dispersion of rounded globules (the white particles in the image) within the brass matrix.

The modification in chip generation mechanism of Pb-bearing brasses is highlighted in the comparative set of micrographs depicted in Fig 6.9. A comparison showing the microstructure and the shape of sectioned chips produced by turning is given for a standard lead-free Cu37Zn alloy and for a free-machining Cu37Zn3Pb brass. It is particularly evident that the presence of Pb (see Fig. 6.9b) promotes the nucleation of numerous cracks that ease the fragmentation of the chip according to the desired mechanism. The amount of Pb in free-machining brass is governed by legal restrictions, but it is also limited by the need to maintain uniform particle dispersion and minimize liquation cracking during hot working processes. As a matter of fact, 4wt.% is temporarily accepted as the maximum Pb level in brasses for general applications, while for parts in contact with potable water the allowed limit until the end of year 2026 is 0.2 wt.%.

6.3 Non-traditional Machining

As shown in Fig. 6.1, the group of non-traditional processes collects all those technologies that remove material without the use of cutting or abrasive tools in physical contact with the workpiece. A variety of alternative methods based on different energy transfer strategies are considered within this broad group. As examples, we can list that mechanical energy is adopted for water jet machining, abrasive jet machining, ultrasonic machining; thermal energy is mainly involved in electro discharge machining, laser jet machining, electron beam machining, whereas chemical and electrochemical energies are used for chemical milling, photochemical milling and electrochemical machining, A representative example of non-traditional process, yet quite widespread in many industrial sectors for high precision products, is electro discharge machining (EDM). Within the topics and discussion covered in this chapter, EDM lends itself as a good example to highlight the effects of machining operation on material microstructure.

In EDM, more precisely called die-sinking EDM, the workpiece and a shaped electrode (the tool) are immersed in a dielectric fluid, usually kerosene or deionized water, and connected to a power source. As a suitable electrical potential between the two parts is reached, the breakdown of the dielectric occurs at different positions within the small gap left between the tool and the workpiece. A series of tiny sparks are produced and a significant rise of temperature and pressure will locally be reached due to the high energy density involved at the spark position. As a result, a little amount of the workpiece material will melt and evaporate. Concurrently the so-formed debris are continuously washed out by the flow of the dielectric fluid, thus freeing the electrode gap and enabling the continuous and controlled erosion of the workpiece. The schematic structure of a EDM system is depicted in Fig. 6.10a. It is assumed that during EDM machining, sparks will form randomly at a relatively large number of positions of the workpiece-tool gap, at a rate of the order of thousands of sparks per seconds. Since this gap will increase as the material is progressively eroded, a control system should let the electrode move lower to maintain a stable process (Kumara, 2023; Sarala Rubi, 2024).

Wire EDM (W-EDM) is a variant of the standard EDM where the electrode assumes the shape of a travelling wire that can cut intricate shapes with high precision. The tiny wire, whose typical diameter is in the range 20–50 μm, is pulled through the workpiece thickness by a supply spoon and collected on the opposite side in a recovery spoon, as depicted in Fig. 6.10b.

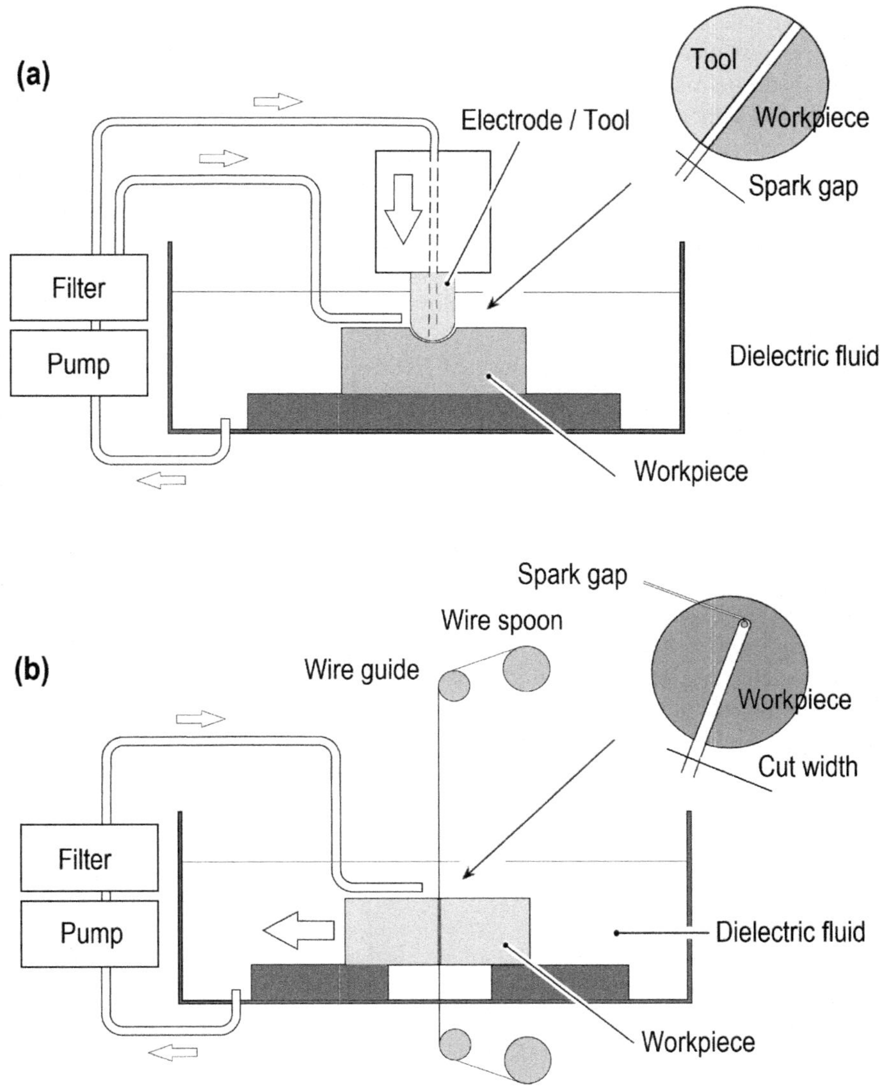

Fig. 6.10 Architecture of EDM (**a**) and W-EDM (**b**) cutting systems (adapted from Kumara & Guptab, 2023)

Given the described erosion mechanisms for EDM and W-EDM, it is expected that the workpiece material would be subjected to substantial microstructural changes due to the sharp thermal gradients generated upon spark formation. For instance, it has been stated (Sidhom, 2013) that each produced spark acts as a heat source increasing the local temperature up to 10,000 °C for a very short time of few microseconds. As a result, a so-called white layer - or recast layer - is often observed on the surface of the workpiece. This layer exhibits characteristics of rapidly solidified material (which may reach an amorphous state under certain conditions) and contains significant residual stresses. Beneath this layer, a thin heat affected zone may become evident in some rough EDM-processed alloys that are more prone to secondary phase precipitation and grain coarsening.

Therefore, the surface morphology of EDM- and W-EDM-processed materials under rough machining parameters can show peculiar features, as displayed in Fig. 6.11a for a 4130 low-alloy steel. The effects left by the material melting and re-solidification induced by the distinct sparks are clearly detectable observing the partially overlapped craters, the droplets adhering on the recast layer and the cavities at the edges of adjacent craters. Figure 6.11b provides direct evidence of the microstructural changes close to surface as detected in sectioned samples. Here, a rough W-EDM cut followed by a set of trim cuts aimed at progressively reducing roughness and material alteration were performed on X38CrMoV5-1 (1.2343) alloy steel samples. It is noticed that a white layer (with reduced reactivity to the metallographic etchant) featuring a thickness of few dozens of micrometres is initially produced by the rough cut. However, this brittle layer can be progressively thinned and almost eliminated by the subsequent trim cuts. For these samples, the microstructural changes beneath the white layer are hardly noticeable since only a faint evidence of carbide dissolution is detected in the heat affected zone, at short distance from the white layer.

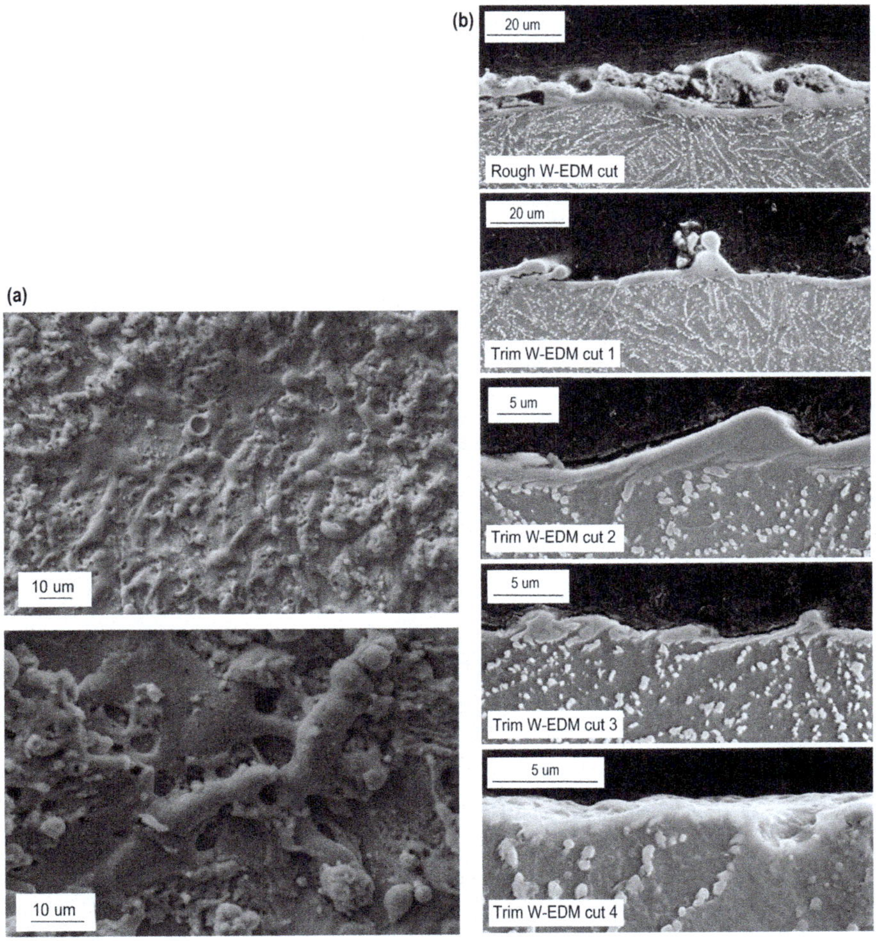

Fig. 6.11 Surface morphology of a W-EDM processed 4130 low-alloy steel (**a**) and microstructural changes after the same type of processing in sectioned samples of a X38CrMoV5-1 alloy steel (**b**)

References

Davis, J. R. (1998). *ASM handbook desk edition.* Machining. ASM International.

Ezugwu, E. O., Bonney, J., & Yamane, Y. (2003). An overview of the machinability of aeroengine alloys. *Journal of Materials Processing Technology, 134*, 233–253.

Kumara, S., & Guptab, T. (2023) A review of electrical discharge machining (EDM) and its optimization techniques. *Materials Today: Proceedings*, 257356734.

Sarala Rubi, C., Prakash, J. U., Juliyana, S. J., Čep, R., Salunkhe, S., Kouril, K., & Ramdas, G. S. (2024). Comprehensive review on wire electrical discharge machining: A non-traditional material removal process. *Frontiers in Mechanical Engineering, 10*, 1322605.

Schultheiss, F., Johansson, D., Bushlya, V., Zhou, J., Nilsson, K., & Ståhl, J. E. (2017). Comparative study on the machinability of lead-free brass. *Journal of Cleaner Production, 149*, 366–377.

Sidhom, H., Ghanem, F., Amadou, T., Gonzalez, G., & Braham, C. (2013). Effect of electro discharge machining (EDM) on the AISI316L SS white layer microstructure and corrosion resistance. *International Journal of Advanced Manufacturing Technologies, 65*, 141–153.

Stavroulakis, P., Toulfatzis, A. I., Pantazopoulos, G. A., & Paipetis, A. S. (2022). Machinable leaded and eco-friendly brass alloys for high performance manufacturing processes: A critical review. *Metals, 12*(246), 1–31.

Zhao, Z., To, S., Wang, J., Zhang, G., & Weng, Z. (2022). A review of micro/nanostructure effects on the machining of metallic materials. *Materials and Design, 224*, Article 111315.

The manufacturer's authorised representative in the EU is Springer Nature Customer Service Centre GmbH, Europaplatz 3, 69115 Heidelberg, Germany. If you have any concerns regarding our products, please contact ProductSafety@springernature.com

Printed and bound by CPI Group (UK) Ltd, Croydon, CR0 4YY
01/10/2026
02225883-0005